Biology, Cultivation and Applications of Capsicum

Biology, Cultivation and Applications of Capsicum

Editor: Allie Duncan

Callisto Reference
www.callistoreference.com

Callisto Reference,
118-35 Queens Blvd., Suite 400,
Forest Hills, NY 11375, USA

Visit us on the World Wide Web at:
www.callistoreference.com

ISBN: 978-1-64116-741-3 (Hardback)

Cataloging-in-publication Data

Biology, cultivation and applications of capsicum / edited by Allie Duncan.
p. cm.
Includes bibliographical references and index.
ISBN 978-1-64116-741-3
1. Peppers. 2. Plants, Protection of. 3. Agriculture. I. Duncan, Allie.
SB351.P4 B56 2023
635.643--dc23

Table of Contents

Preface

Capsicum refers to a genus of flowering plants in the nightshade family solanaceae. It is farmed for its bell pepper fruit or chili pepper throughout the world. Capsicum is a major spice crop and vegetable, which is grown for dried, processed and fresh products. It is primarily a temperate crop that requires a low temperature for fruit set and its growth. The low temperature is also responsible for the color development of capsicum. It thrives during summer in the hills and in the plains during winters. Capsicum comes in a variety of species and hues and can taste tart and sweet; however, green variants have a little more flavor. It has a wide range of flavors and nutritional qualities. There are several benefits of capsicum such as reducing anxiety, improving bone health, preventing anemia and cancer, boosting immunity, and improving eye health. This book unravels the biology, cultivation and applications of capsicum. It will serve as a valuable source of reference for graduate and postgraduate students.

This book is a comprehensive compilation of works of different researchers from varied parts of the world. It includes valuable experiences of the researchers with the sole objective of providing the readers (learners) with a proper knowledge of the concerned field. This book will be beneficial in evoking inspiration and enhancing the knowledge of the interested readers.

In the end, I would like to extend my heartiest thanks to the authors who worked with great determination on their chapters. I also appreciate the publisher's support in the course of the book. I would also like to deeply acknowledge my family who stood by me as a source of inspiration during the project.

Editor

1

Management of Viruses and Viral Diseases of Pepper (*Capsicum* spp.) in Africa

Olawale Arogundade, Titilayo Ajose, Itinu Osijo, Hilary Onyeanusi, Joshua Matthew and Taye H. Aliyu

Abstract

Increasing outbreaks of virus species infecting pepper (*Capsicum* spp.) is a major problem for growers in Africa due to a combination of factors, including expansion of pepper cultivation, abundance of insect vectors and climate change. More than 45 viruses have been identified to infect pepper crops causing economic loss in terms of reduced quality and marketable yield, sometimes up to 100%. The *Pepper veinal mottle virus* (PVMV), *Potato virus Y* (PVY) and *Cucumber mosaic virus* (CMV) are endemic in many countries including Uganda, Mali, Cameroon, Morocco and Nigeria. Current management options for virus infection in *Capsicum* spp. is by the integration of several approaches. More importantly, eradication of infected plants, cultivation of disease resistant varieties, improved cultural practices and judicious use of insecticides especially when plants are young and easily colonized by vectors. In recent years, eco-friendly control measures are needful to reduce occurrence of virus diseases in *Capsicum* spp.

Keywords: climate change, economic loss, outbreaks, management options, virus infection

1. Introduction

Peppers (*Capsicum* spp.) are one of the most important spices and vegetable crops in the economic and social life of people living worldwide [1]. Viruses are among the most important factors threatening *Capsicum* spp. production in several regions like Australia [2], Europe [3], Asia [4] and Africa [5]. They cause diseases that not only reduce yield and quality of fruits, but also increase the cost of preventive measures and cost of producing clean planting materials. In addition, the high genetic diversity of virus strains and their accumulation in propagation materials makes them easily spread into unaffected areas [6].

In Africa, more than 45 viruses (grouped into eight genera namely, *Potyvirus*, *Tospovirus*, *Begomovirus*, *Cucumovirus*, *Tobamovirus*, *Polerovirus*, *Alfamovirus* and *Potexvirus*) have been identified to infect pepper crops causing reduced quality and marketable yield, sometimes up to 100% [7]. The major viruses of *Capsicum* spp. include *Pepper veinal mottle virus* (PVMV), *Chilli veinal mottle virus* (ChiVMV), *Potato virus Y* (PVY), *Tobacco etch virus* (TEV), *Tobacco mosaic virus* (TMV), *Tomato mosaic virus* (ToMV), *Pepper mild mottle virus* (PMMoV), *Pepper leaf curl virus*

(PepLCV), *Tomato yellow leaf curl virus* (TYLCV), *Cucumber mosaic virus* (CMV), *Alfalfa mosaic virus* (AMV) and *Tomato spotted wilt virus* (TSWV) while other viruses of minor importance include *Pepper vein yellows virus* (PeVYV) and *Potato virus X* (PVX) [8]. These viruses are mainly vectored by aphids, whitefly or thrips during feeding probes. However, mechanical transmission and seed transmission are efficient means for the spread of infection. Typical symptoms induced by viruses infecting *Capsicum* spp. include stunting, curling and mottling of foliage and fruits. Symptoms can vary significantly with cultivar, plant age, virus isolate and environmental condition. In recent years, co-infection of previously non-existent virus strains is most devastating [7].

The increasing outbreaks of virus species infecting *Capsicum* spp. have become a major problem for growers in many countries including Ghana, Uganda, Mali, Cameroon, Morocco and Nigeria [9, 10]. This is due to a combination of factors, including expansion and intensification of pepper cultivation, availability of volunteer hosts, abundance of insect vectors and climate change [6, 11]. Many techniques are used to minimize viral infections in *Capsicum* spp., but are successful when they begin prior to planting of the crop. Current management options for virus infection in *Capsicum* spp. is by the integration of several approaches. These include use of protected nurseries, cultivation of disease resistant varieties and ensuring adequate phyto-sanitary conditions after transplanting.

Generally, eco-friendly management measures are needful to reduce occurrence of virus diseases in *Capsicum* spp. and decrease the rate of spread of same into unaffected regions. This is important because of the need for increased international movement and exchange of pepper germplasm. This will also facilitate the selection and breeding for improved *Capsicum* spp. adapted to local conditions and uses. Therefore, there is need for a review of viruses infecting *Capsicum* spp., their symptoms, mode of transmission and management options.

2. Viruses of pepper

2.1 Genus: *Potyvirus*

The genus *Potyvirus* consists of a large group of plant viruses that are widespread, infecting economically important crops like tomato, potato, eggplant and pepper [12]. Members of this genus have non-enveloped flexuous filamentous single-stranded RNA (ssRNA) particles about 680–900 nm long and 11–15 nm wide [13]. The symptoms induced by potyviruses include mottling, mosaic, curling, vein banding or clearing, chlorosis, blistering and severe stunting of the whole plant. Severity of these symptoms depends on the *Capsicum* variety, virus strain and presence of other viruses. The major potyviruses infecting *Capsicum* spp. in Africa include PVMV, ChiVMV, PepMoV, PVY and TEV [12].

2.1.1 Pepper veinal mottle virus *(PVMV)*

Pepper veinal mottle virus (PVMV) was first reported in Ghana in 1971 and has since then spread to other regions in Africa including Ethiopia, Kenya, Tanzania, Uganda and Rwanda [14] causing yield loss ranging from 70 to 100 % [15]. Infection of PVMV also occurs in other solanaceous crops, including eggplant and tomato [16]. Eight species of aphids have been shown to transmit the virus in a non-persis-tent manner, of which *Myzus persicae*, *Aphis craccivora*, *A. gossypii* and *A. spiraecola* are rated as efficient vectors [17]. Mechanical transmission also occurs through contact with infected sap, but not via seeds. Several symptoms associated with

PVMV include leaf mottle, leaf mosaic, leaf curl, vein banding, leaf ring spots, leaf deformation, leaf chlorosis, blistering and severe stunting of the whole plant [18].

2.1.1.1 Management of PVMV

Plastic mulches have been reported to reduce insect infestation and the concurrent damage associated with the insect transmission of viruses [10]. Intercropping *Capsicum* spp. with maize can also reduce the incidence of the disease compared to sole cropping [17]. The use of resistant varieties, raising of seedlings in insect-proof nursery and roguing of infected plants once detected are also crucial to reduce PVMV incidence and spread in the field. These are recommended as integrated ways for managing PVMV infection in the field [7].

2.1.2 Chilli veinal mottle virus *(ChiVMV)*

Chilli veinal mottle virus (ChiVMV) has been reported in various countries of West and East Africa including Ethiopia, Uganda and Tanzania [19]. The virus is not transmitted through seeds but is acquired mechanically and is transmitted by several species of aphids in a non-persistent manner. These vectors (*Myzus persicae*, *Aphis gossypii*, *Aphis craccivora*, *Aphis spiraecola*, *Rhopalosiphum maidis*, *Toxoptera citricida* and *Hysteroneura setariae*) retain the virus for not more than1 hour after virus acqui-sition [20]. The winged aphids are generally the most efficient vectors to transmit the virus from field to field and are the most difficult to control [21]. Typical symptoms caused by ChiVMV include leaf mottle and dark green vein-banding.

2.1.2.1 Management of ChiVMV

The effective management strategies for ChiVMV include early identifica-tion and timely disposal of infected seedlings, application of mineral oil-based insecticides in controlling aphid populations early enough during planting, care-ful handling of diseased-free seedlings and use of resistant varieties. In addition, removal of solanaceous weeds, which serve as alternate hosts, can reduce incidence of ChiVMV in pepper gardens [21].

2.1.3 Potato virus Y *(PVY)*

Potato virus Y (PVY) is common among solanaceous crops including pepper, potato, tomato, tobacco and many weeds [22]. It has been reported in many African countries including Zimbabwe, Zambia, Kenya, Tanzania, Malawi, Madagascar, Ethiopia and Nigeria [23]. Isolates of PVY from pepper do not infect potato and vice versa [24]. Strains of PVY infecting sweet pepper are classified by their ability to overcome resis-tance genes based on differential host reactions. They cause yield reductions of 20–70% in pepper production [25]. The virus is transmitted in a non-persistent manner through several species of aphids, but the green peach aphid (*Myzus persicae*) is considered generally to be the most important vector. Infection of PVY can spread during grafting, handling of plant and use of unsterilized farming equipment. The symptoms of PVY include stunting or dwarfing of plant, systemic vein clearing and banding, leaf mosaic and small deformed fruit with a mosaic pattern making them unmarketable [26].

2.1.3.1 Management of PVY

Management of PVY is successful through the use of resistant varieties, elimination of solanaceous weeds and adjoining volunteer host plants, scheduling the time of planting to conform with period of least aphid population, careful handling of plant seedlings especially during nursery and farming operations. Early detection and roguing of infected plants are the effective measures to reduce incidence of PVY. Additionally, intercropping pepper with maize has been found helpful in the management of potyviruses in pepper fields [27].

2.1.4 Tobacco etch virus *(TEV)*

Tobacco etch virus (TEV) naturally occurs in co-infection with PVY [13]. In Africa, TEV is endemic and widespread in Madagascar and occasionally in South Africa. The virus is transmitted by several species of aphids in a non-persistent manner. Infected plants are lighter in colour and have mosaic patterns on their leaves, sometimes with dark-green banding along the leaf veins. Plants may also be stunted, showing leaf curl and fruit distortion.

2.1.4.1 Management of TEV

The use of cultivars with PVY resistance helps manage TEV because resistance to the two viruses is closely-linked. Additionally, pepper seedlings should be produced in protected culture system equipped with nets or screens to exclude aphids. Scheduling planting dates to avoid periods of high aphid activity early in the season and controlling weeds, volunteer host plants in and around seedlings will lower the amount of virus inoculum introduced into the crop. Application of mineral oil has been reported to reduce incidence of TEV in pepper [10].

2.2 Genus: *Tobamovirus*

The viruses in the genus *Tobamovirus* are especially important because they do not require biological vectors for transmission. Members of this genus have flexuous filamentous ssRNA particles approximately 300 nm in length and 18 nm in diam-eter that can persist for years and survive many adverse environmental conditions because of their strong structural coat protein. Due to their high stability, tobamoviruses remain infectious in contaminated plant debris, compost, soil and irrigation water. Viruses in this genus include TMV, ToMV and PMMoV [13].

2.2.1 Tobacco mosaic virus *(TMV)*

Tobacco mosaic virus (TMV) is the first ever virus to be identified. The virus infects more than 350 plant species, including at least 125 crop species such as tobacco, tomato, pepper, eggplant, potato and cucumber [28]. *Tobacco mosaic virus* has been found to infect pepper in Uganda, Tanzania, Zimbabwe, Sudan, Zambia (East Africa) and (Nigeria, Ghana) West Africa [29, 30]. The virus survives in infected plant materials for months or years and multiplies in living plant tissue but can remain dormant in dead plant tissue, retaining its infectiveness [31]. Tobacco mosaic virus spreads mainly through contact between plants, infected seeds and by mechanical means but not insect vectors. Typical symptoms of TMV infection include chlorotic leaves, mosaic patterns on leaves, leaf distortion and stunted growth usually associated with reduced fruit size [32].

2.2.1.1 Management of TMV

Management of TMV can be achieved through careful handling of plants, disinfection of farm tools, eradication of infected plants and treatment of seeds with 10% trisodium phosphate (TSP). The use of resistant cultivar to TMV infection has been demonstrated in Nigeria [33]. Also, healthy seed-testing and 2-year minimum crop rotation is advisable. *Capsicum* spp. should not be planted alongside other susceptible crops such as tomato and potato.

2.2.2 Tomato mosaic virus *(ToMV)*

Tomato mosaic virus (ToMV) has a worldwide distribution and is often endemic in African countries such as Uganda, Zambia and Cameroon [29]. The virus is one of the most resilient viruses in terms of its ability to survive outside plant cells and dead tissues. *Tomato mosaic virus* is known to infect more than 150 economically important crop species, including vegetables and ornamental plants. Natural mode of spread of ToMV is by contact with infected sap but it is also being transmitted mechanically during farming operations [6]. The occurrence of ToMV is more predominant in pepper than TMV even though both virus species produce similar symptoms on pepper. Typical symptoms include severe stunting of plants with chlorotic mosaic patterns on leaves and fruits [34]. Additional symptoms may vary depending on the plant age, virus strain and environmental condition. These include distorted leaves, which often occur with premature defoliation and necrotic (brown) patches on leaves and fruits [35].

2.2.2.1 Management of ToMV

Management options for ToMV include rotation to non-host plants after an infection outbreak. The use of resistant varieties coupled with improved cultural practices can help to improve production especially in endemic areas. Disinfection of screen-house soils, planting tools and containers ensures clean planting materials in the nursery. After transplanting, careful handling of healthy seedlings while removing symptomatic ones can reduce spread of the virus. Sterilizing seeds with 10% TSP can help remove ToMV present on the seed coat [36].

2.2.3 Pepper mild mottle virus *(PMMoV)*

Pepper mild mottle virus (PMMoV) has been widely reported in some African countries including Uganda, Zambia, Tanzania, Nigeria and Ghana. It has been shown to infect up to 24 other plant species including *Solanaceae*, *Chenopodiaceae*, *Cucurbitaceae*, *Labiatae* and *Plantaginaceae* [29, 23]. In cultivated pepper plants, PMMoV can be transmitted through seed and contact with infected plant sap [37]. Seedlings can also be infected by mechanical contamination during transplanting or other cultural routine. *Pepper mild mottle virus* persists in soil and on infected debris, which serve as primary source of inoculum for subsequent planting. Symptoms of PMMoV include various degrees of leaf mottling, leaf mosaic, leaf chlorosis, necrosis, leaf curl and growth decline. These symptoms are more pronounced in young plants than in older plants [38].

2.2.3.1 Management of PMMoV

Treatment of *Capsicum* spp. seeds with 10% TSP for 2.5 hours (h) reduces the incidence of PMMoV. Other management options include sterilization of soil before planting, especially in screen-houses and ensuring good field hygiene after trans-planting [13].

2.3 Genus: *Begomovirus*

The genus *Begomovirus* comprises monopartite or bipartite plant viruses that infect a wide range of crops throughout the world. Viruses in this genus are exclusively transmitted by the whitefly (*Bemisia tabaci*) in a persistent and circulative manner. These virus species have circular single-stranded DNA (ssDNAs) particles approximately 30 nm in length and 18 nm in diameter. Examples of important begomoviruses affecting pepper include PepLCV and TYLCV.

2.3.1 Pepper leaf curl virus *(PepLCV)*

Pepper leaf curl virus (PepLCV) has been found in different countries of Africa and is responsible for several epidemics causing severe economic losses. Transmission of PepLCV is most damaging when plants are infected at early growth stage preventing the proper formation of flowers resulting in low fruit production. Symptoms of PepLCV include severe stunting, flower bud abscission, reduced pollen production, upward curling of leaves, leaf chlorosis and elimination of fruit production. Yield losses may be between 90 and 100% [39].

2.3.1.1 Management of PepLCV

The most widely used treatments include use of insecticides and other cultural methods to control vector population. Other methods to control the spread of PepLCV include planting resistant or tolerant plants, crop rotation and border planting, and plastic mulching [40].

2.3.2 Tomato yellow leaf curl virus *(TYLCV)*

Tomato yellow leaf curl virus (TYLCV) is *one* of the most destructive viruses affecting a wide host range of vegetable crops, including Okra (*Abelmoschus esculentus*), tomato (*Solanum lycopersicum*), sweet and chilli pepper (*Capsicum* spp.), tobacco (*Nicotiana tabacum*), common bean (*Phaseolus vulgaris*) and some weeds [41]. The virus has been reported in *C. annuum* in Tunisia [42]. However, its distribution spreads across East, West and Central Africa. Movement of infected plants or virus-carrying whiteflies aids the spread of the virus. Typical symptoms of TYLCV in infected pepper plants include interveinal yellowing, leaf curling and stunting which cause critical production losses.

2.3.2.1 Management of TYLCV

Plastic mulching has been effective in the control of whitefly populations in pepper fields. Other strategies for managing TYLCV include raising of seedlings in pest-proof nurseries, adoption of crop rotation, use of insecticides, selection of resistant varieties for planting, improved cultural practices to remove weeds and alternative host plants [43].

2.4 Genus: *Cucumovirus*

The genus *Cucumovirus* comprises tripartite ssRNAs encapsidated in small icosahedral particles approximately 29 nm in diameter. Viruses in this genus infect over 1200 plant species worldwide [44], including weeds and wild species.

2.4.1 Cucumber mosaic virus *(CMV)*

Cucumber mosaic virus (CMV) is an important viral disease of pepper with a worldwide distribution. In Africa, the virus has been reported in Uganda, Zambia,

Tanzania, Ethiopia, Zimbabwe, Kenya, Malawi, Madagascar, Sudan, Rwanda, Ghana and Nigeria [14, 45]. Although the virus exists as a number of strains, all are apparently capable of infecting pepper and differ only in their symptom expression. The age of a plant at the time of infection strongly influences what type of symptoms will be manifested. Symptoms of naturally affected pepper plants vary and most prominent ones are mild mosaic and dull-coloured leaves, mottling, shoe string, fern leaf, vein banding, vein clearing, leaf deformation, stunted growth and reduced fruit size [46, 47]. More than 80 species of aphids transmit the virus in a non-persistent way but *Aphis gossypii* and *Myzus persicae* are the most efficient [48]. Additionally, transmission can be through seeds, parasitic weeds and mechanically [26, 49].

2.4.1.1 Management of CMV

Strategies to delay early infection can be used to enhance yield. Isolation of pepper from weedy border areas or growing them next to taller border plants such as maize, which can function as a non-susceptible host, use of certified seeds and plants, screening and disinfection of infected mother stock, washing and disinfecting of hands and tools, and planting of resistant pepper genotypes constitute the effective means of managing CMV [47].

2.5 Genus: *Alfamovirus*

The genus *Alfamovirus* consists of ssRNA viral particles. A typical member of this genus is *Alfalfa mosaic virus* (AMV) having natural host range including over 250 plant species and is closely related to CMV [13].

2.5.1 Alfalfa mosaic virus *(AMV)*

Alfalfa mosaic virus (AMV) is distributed worldwide, having a wide host range. In Africa, it has been reported in Zambia [50]. The virus is transmitted mechanically and in a non-persistent manner by numerous aphid species. Transmission of AMV also occurs through pepper seeds or pollen [9]. Typical symptoms associated with AMV in pepper include bright yellow or blotchy white mosaic patterns on pepper leaves [6]. Additional symptoms include stunted growth with deformed and blotchy fruits, especially if plants are infected at young stage. *Alfalfa mosaic virus* can cause important yield losses and increased susceptibility of pepper to other pathogens.

2.5.1.1 Management of AMV

Alfalfa mosaic virus can be successfully managed by reducing aphid population by use of insecticides. The use of resistant varieties and regular rouging of weeds from the fields are effective means to manage the virus [51].

2.6 Genus: *Tospovirus*

The genus *Tospovirus* consists of important species of plant viruses that cause great losses in many economically important crops. Members of this genus have tripartite ssRNA particles about 80–120 nm in size [42, 52]. The major virus species in the genus *Tospovirus* is *Tomato spotted wilt virus* (TSWV), which infects various *Capsicum* spp. in Africa [13].

2.6.1 Tomato spotted wilt virus *(TSWV)*

Tomato spotted wilt virus (TSWV) causes serious crop losses in many economically important crops, including vegetables and ornamental crops. It has been reported to affect pepper in Zimbabwe in Africa [53]. The virus is transmitted in a persistent and propagative manner by several species of thrips. The western flower thrip (*Frankliniella occidentalis*) is the most efficient [54]. However, only the adult thrips that feed on infected plants transmit the virus after inoculation period of less than 48 h. *Tomato spotted wilt virus* is known to infect more than 1000 different plant species from about 80 botanical families. Symptoms of the virus can be very host-specific. Typical symptoms observed in pepper include yellowing or browning of leaves, chlorotic or necrotic ringspots on leaves and fruits, necrotic streaks on stems with terminated shoots and fruits [6].

2.6.1.1 Management of TSWV

Early detection and roguing of infected plants is important for reducing TSWV incidence in the field. Intercropping pepper with companion crops like ginger (*Zingiber officinale*) can reduce the vector from reaching its host. Additionally, use of insecticide is a critical measure in management of TSWV [9].

2.7 Genus: *Polerovirus*

2.7.1 Pepper vein yellows virus *(PeVYV)*

Pepper vein yellows virus (PeVYV) has been reported in Ivory Coast, Mali, Republic of Benin, Sudan and Tunisia in Africa with infection rates of up to 100%. The virus is spread in circulative and non-propagative manner by *A. gossypii* and *M. persicae* [55]. Major host plants include *Capsicum* spp. and *Solanum nigrum* [56]; however, some alternate hosts such as *Chenopodium amaranticolor*, *Curcubrita pepo*, *Datura stramonium*, *Gomphrena globose* and *Nicotiana* spp. have been reported [56]. Symptoms on cultivated pepper plants include leaf curling, deformation, reduced leaf size, puckering, interveinal yellowing, vein clearing and yellow patches on leaves [57].

2.7.1.1 Management of PeVYV

The most effective means of controlling PeVYV is preventing its introduction. Careful sourcing of plants, keeping aphid population low, raising awareness of its symptoms will assist in preventing the establishment of the virus [56].

2.8 Genus: *Potexvirus*

2.8.1 Potato virus X *(PVX)*

Potato virus X (PVX) has been reported in Ethiopia, Zambia and Zimbabwe to infect a wide range of solanaceous crops including pepper, tomato, potato and tobacco. The virus causes a range of symptoms including mottling, severe necrosis of leaves and stems and sometimes, defoliation of some cultivars. Symptoms of PVX are worsened in co-infection with other viruses, especially *Potato virus Y* [58, 59]. *Potato virus X* is transmitted mechanically and by contact between plants, but not through seeds [60]. PVX is of minor importance in pepper production [60].

2.8.1.1 Management of PVX

Planting resistant cultivar is the most economic and effective way of managing the virus [61].

3. Conclusion

Viruses remain a primary constraint to production of high-quality *Capsicum* spp. worldwide, especially in the developing regions [1]. In Africa, weeds often become reservoir hosts for vectors and virus species that attack pepper in the field [47]. Currently, virus species in the genus *Potyvirus*, *Cucumovirus*, *Begomovirus* and *Tobamovirus* are major concern for many growers because of their effects on yield [7, 62]. The severity of infection depends on environmental conditions, host varieties and individual virus infection (**Figure 1**). Generally, these virus infections cannot be totally eradicated in many plantations where they occurred [10]. However, prompt action against the damage caused by viruses is with the use of resistant varieties and ensuring adequate phyto-sanitary conditions within the field. Therefore, awareness of local farmers on the impact of field hygiene must be improved as adjunct to using tolerant varieties. Additionally, screening of young seedlings for infection before they ever reach the field is crucial to reduce virus occurrence. Ultimately, the development of eco-friendly ways of virus disease management will help to improve yield in the pepper industry.

Figure 1.
Possible symptoms that can be observe on pepper in the field. (A) Mosaic pattern on leaves, (B) leaf mottling, (C) leaf curl, (D) vein banding, (E) vein yellowing and (F) leaf reduction.

Author details

Olawale Arogundade[1*], Titilayo Ajose[1], Itinu Osijo[1], Hilary Onyeanusi[1], Joshua Matthew[1] and Taye H. Aliyu[2]

1 National Horticultural Research Institute (NIHORT), Ibadan, Oyo State, Nigeria

2 Department of Crop Protection, University of Ilorin, Ilorin, Kwara State, Nigeria

*Address all correspondence to: arogundade_olawale@yahoo.co.uk

References

[1] Prabhakaran-Nair K, P. The agronomy and economy of black pepper (*Piper nigrum* L.): The king of spices. Advances in Agronomy. 2011;**82**:1-108

[2] Maina S, Edward OR, Jones RAC. First complete genome sequence *of Pepper vein yellows virus* from Australia. Genome Announcement. 2016;**4**(3):450-416

[3] Svoboda J, Svobodova-Leisova L. Occurrence of viruses on pepper plantations in the Czech Republic-Short Communication. Horticultural Science (Prague). 2012;**39**:139-143

[4] Amisa L, Somrudee N, Ekachai C. Incidence of viruses infecting pepper in Thailand. Biological and Molecular Concepts. 2019;**10**:184-193

[5] Aliyu TH. The incidence, severity and occurrence of four viruses infecting pepper (*Capsicum* spp.) in the Southern Guinea Savannah Agro-ecological Zone of Nigeria. Agriculturae Conspectus Scientifius. 2014;**79**:233-237

[6] Kenyon L, Kumar S, Wen-Shi Tsai, Jacqueline d'A H. Virus diseases of peppers (*Capsicum* spp.) and their control. Advances in Virus Research. 2014;**90**:297-354

[7] Waweru BW, Kilalo DC, Miano DW, Kimenju JW, Rukundo P. Diversity and economic importance of viral diseases of pepper (*Capsicum* spp.) in Eastern Africa. Journal of Applied Horticulture. 2019;**21**(1):70-76

[8] Arogundade O, Kareem KT, Lava KP, Omodele T. Prevalence of *viruses* in field grown pepper in Oyo and Osun states of Nigeria. Nigerian Journal of Plant Protection. 2014;**28**:117-125

[9] AVRDC. Tomato spotted wilt virus. Fact Sheet. AVRDC Publication 05-626. 2005. Available from: https://www.google.com/url?q=http://avrdc.org/%3Fwpfb_dl%3D202&sa=U&ved=2ahUKEwiq9pKJiZXoAhWxz4UKHfpCBgsQFjABegQlCBAB&usg=AOvVaw0kCBVTILE8pEaFHu0L7Cel [Accessed: 21 February 2020]

[10] Bosland PW, Votata EJ. Peppers: Vegetable and Spice Capsicums. 2nd ed. London: CAB International Institute; 2012. 217 pp. ISBN: 9781845938253

[11] German TL, Ullman DE, Moyer JW. *Tospoviruses*: Diagnosis, molecular biology, phylogeny and vector relationships. Annual Review Phytopathology. 1992;**30**:315-348

[12] Roberts PD, Adkins S, Pernezny K, Jonnes JB. Diseases of pepper and their management. In: Naqvi SAMH, editor. Diseases of Fruits and Vegetables. Vol. 11. Kluwer Academic; 2004. pp. 333-387

[13] Moury B, Verdin E. Viruses of pepper crops in the Mediterranean basin: A remarkable stasis. Advances in Virus Research. 2012;**86**:127-162

[14] Skelton A, Uzayisenga B, Fowkes A, Adams I, Buxton-Kirk A, Harju V, et al. First report of *Pepper veinal mottle virus, Pepper yellows virus* and a novel *enamovirus* in chilli pepper (*Capsicum* spp.) in Rwanda. New Disease Reports. 2018;**37**:5

[15] Fajinmi AA, Popoola AR, Fajinmi OB. Incidence and epidemiology of *Pepper veinal mottle virus* (PVMV), a Potyvirus disease of pepper. Archives of Phytopathology and Plant Protection. 2012;**45**(13):1586-1590

[16] Alegbejo MD, Abo ME. Ecology, epidemiology and control of *Pepper veinal mottle virus* (PVMV) genus *Potyvirus*, in West Africa. Journal of Sustainable Agriculture. 2002;**20**(2):5-16

[17] Fajinmi AA, Odebode CA. Evaluation of maize/pepper intercropping model in the management of *Pepper veinal mottle virus*, genus *Potyvirus*, family Potyviridae on cultivated pepper (*Capsicum annuum* L.) in Nigeria. Archives of Phytopathology and Plant Protection. 2010;**43**(15):1524-1533

[18] Tsai WS, Abdourhamane IK, Kenyon L. First report of pepper veinal mottle virus associated with mosaic and mottle diseases of tomato and pepper in Mali. Plant Disease. 2010;**94**:378-378

[19] Womdim NR, Swai IS, Chadha ML, Selassie GK, Marchoux G. Occurrence of *chilli veinal mottle virus* in *Solanum aethiopicum* in Tanzania. Plant Disease. 2001;**85**:801

[20] Shah H, Yasmin T, Fahim M, Hameed S, UI-Haque, M. I. Transmission and host range studies of Pakistani Isolate of *Chilli veinal mottle virus*. Pakistan Journal of Botany. 2008;**40**:2669-2681

[21] Arade PC. Serological detection, molecular characterization and management of *Chilli veinal mottle virus* (ChiVMV) in chilli (*Capsicum* spp.) [thesis]. Department of Plant Pathology: Anand Agricultural University; 2015. 109 pp

[22] Ibaba JD, Gubba A. Phylogenetic studies of selected isolates of *Potato virus Y* (PVY) infecting vegetable crops in KwaZulu-Natal, Republic of South Africa. South African Journal of Plant and Soil. 2012;**29**(2):117-120

[23] Olawale A, Olusegun SB, Kehinde TK. Occurrence and distribution of *Pepper veinal mottle virus* and *Cucumber mosaic virus* in pepper in Ibadan. Nigeria Virology Journal. 2012;**9**:79

[24] Singh RP, Valkonen JPT, Gray SM, Boonham N, Jones RAC, Kerlan C, et al. The naming of *Potato virus Y* strains infecting potato. Archives of Virology. 2008;**153**:1-13

[25] Tsedaley B. A review paper on *Potato virus Y* (PVY) biology, economic importance and its managements. Journal of Biology, Agriculture and Healthcare. 2015;**5**(9):110-126

[26] Arogundade O, Balogun OS, Kumar LP. Seed transmissibility of *Cucumber mosaic virus* in *Capsicum* spp. International Journal of Vegetable Science. 2019;**25**(2):146-153

[27] Mitiku A, Chala A, Beyene Y. The effect of intercropping pepper with maize and sweet potato on infection of pepper (*Capsicum annuum* L.) by *potyviruses* and yield of pepper in Southern Ethiopia. International Journal of Technology Enhancements and Emerging Engineering Research. 2013;**1**(4):68-73

[28] Kumar S, Udaya AC, Shankar SC, Nayaka OS, Lund, Prakash HS. Detection of Tobacco mosaic virus and Tomato mosaic virus in pepper and tomato by multiplex RT-PCR. Letters in Applied Microbiology. 2011;**53**: 359-363

[29] IPM CRSP. Annual Report. Integrated Pest Management Collaborative Research Support Program. USAID: Virginia Tech; 2008. p. 190

[30] Arogundade O, Ajibona OD, Ogunleti OD, Oke KE. Intercrop and crop combination effect on virus disease incidence, growth and yield of *Capsicum* spp. in a pepper-maize intercrop. Journal of Biological and Chemical Research. 2015;**32**(2):945-952

[31] Damiri N, Sofita IS, Effend TA, Rahim SE. Infection of some cayenne pepper varieties (*Capsicum frutescens* L.) by *Tobacco mosaic virus* at different growth stages. In: 3rd Electronic and Green Materials International

Conference 2017; AIP Conference Proceedings. Vol. 1885(1). American Institute of Physics; 2017. DOI: 10.1063/1.5005942

[32] Pazarlar S, Gumus M, Oztekin GB. The effects of *Tobacco mosaic virus* infection on growth and physiological parameters in some pepper varieties (*Capsicum annuum* L.). Notulae Botanicae Horti Agrobotanici Cluj-Napoca. 2013;**41**(2):427-433

[33] Igwegbe ECK, Ogungbade K. Evaluation of pepper cultivars under greenhouse conditions for resistance to a defoliation strain of *Tobacco mosaic virus*. Plant Disease. 1985;**69**:899-900

[34] Arogundade O, Aderonmu OI, Matthew JO, Ayo-John EI. First report of tomato mosaic virus Isolated from *Solanum macrocarpon* in Nigeria. Plant Disease. 2018;**102**(2):458

[35] Pernezny K, Robert PD, Murphy JF, Goldberg NP. Compendium of Pepper Diseases. Vol. 1. USA: The American Phytopatholgical Society. pp. 24-25

[36] Rast ATB, Stijger CCMM. Disinfection of pepper seed infected with different strains of *capsicum mosaic virus* by trisodium phosphate and heat treatment. Plant Pathology. 1987;**36**:583-588

[37] Nikolay P. Effect of pepper mild mottle virus infection on pepper and tomato plants. Science and Technologies. 2014;**5**:61-64

[38] Sevik M. A occurrence of pepper mild mottle virus in greenhouse grown pepper (*Capsicum annuum* L.) in the West Mediterranean region of Turkey. African Journal of Biotechnology. 2011;**10**(25):4976-4979

[39] Kumar S, Kumar S, Singh M, Singh AK, Rai M. Identification of host plant resistance to *Pepper leaf curl virus* in chilli (*Capsicum* spp.). Scientia Horticulturae. 2006;**110**(4):359-361

[40] Rai VP, Kumar R, Singh SP, Kumar S. Monogenic recessive resistance to *Pepper curl virus* in an interspecific cross of *Capsicum*. Scientia Horticulturae. 2014;**172**:34-38

[41] Diaz-Pendon JA, Canizares MC, Moriones E, Bejarano ER, Czosnek H, Navas-Castillo J. Tomato yellow leaf curl viruses: Menage a trois between the virus complex, the plant and the whitefly vector. Molecular Plant Pathology. 2010;**4**(11):441-450

[42] Gharsallah Chouchane S, Gorsane F, Nakhla MK, Maxwell DP, Marrakchi M, Fakhfakh H. First report of *Tomato yellow leaf curl virus* Isreal species infecting tomato, pepper and bean in Tunisia. Journal of Phytopathology. 2007;**155**:236-240

[43] Antignus Y. The management of Tomato yellow leaf curl virus in Greenhouses and the open field, a strategy of manipulation. In: Czosnek H, editor. *Tomato Yellow Leaf Curl Virus Disease*. Dordrecht: Springer; 2007. 263-278 p. ISBN 978-1-4020-4769-5

[44] Palukaitis P, Garcia-Arenal F. *Cucumovirus*. Advanced Virus Research. 2003;**62**:241-323

[45] Appiah AS, Quartey HE, Amoatey K, Nunekpeku MW, Owusu-Ansah M, Ofori S. Response of nine cultivars of pepper (*Capsicum* spp.) to infection by four viruses under natural field conditions in the coastal savanna zone of Ghana. Research Journal of Applied Sciences, Engineering and Technology. 2014;**7**(5):903-907

[46] Kapoor S, Sharma A, Handa A. Correlation between symptoms and ELISA for the detection of *Cucumber mosaic virus* in bell Pepper. International Journal of Current Microbiology and Applied Sciences. 2018;**7**(6):400-406

[47] Arogundade O, Oresanya A, Matthew JO, Kareem KT, Onyeanusi H.

Natural occurrence of Cucumber mosaic virus in Ogeira (*Eleutheranthera ruderalis*) in Nigeria. Archives of Pathology and Plant Protection. 2019. DOI: 10.1080/03235408.2019.1585407

[48] Palukaitis P, Roossinck MJ, Dietzgen RG, Francki RIB. *Cucumber mosaic virus*. Advances in Virus Research. 1992;**41**:281-349

[49] Akhtar KP, Anwer M, Saleem MY, Wousaf S, Ullah N, Cheema HMN, et al. Identification of natural weed hosts of *Cucumber mosaic virus* subgroup-1 and the absence of seed transmission in weed hosts in Pakistan. Journal of Horticultural Science and Biotechnology. 2019;**94**(4):468-474

[50] Kaitisha GC. Some virus diseases of crop plants in Zambia. In: Hughes J d'A, Odu BO, editors. Proceedings of International Institute of Tropical Agriculture Conference on Plant Virology in Sub Saharan Africa; 4-8 June 2001; Ibadan, Nigeria: 2001. pp. 317-333

[51] Aftab M, Angela F. Temperate pulse viruses. *Alfalfa mosaic virus*. Agriculture Notes. DEPI Field Crops Pathology; 2000. 1-3 pp

[52] Law MD, Speck J, Moyer JW. The mRNA of impatiens necrotic spot *tospovirus* (Bunyaviridae) has an ambisense genomic organisation. Virology. 1992;**188**:732-741

[53] Nyamupingidza TN, Machakaire V. Virus diseases of important vegetables in Zimbabwe. In: Hughes J d'A, Odu BO, editors. Proceedings of International Institute of Tropical Agriculture Conference on *Plant Virology* in Sub Saharan Africa; 4-8 June 2001; Ibadan, Nigeria. 2001. pp. 397-405

[54] Gallitelli D, Mascia T, Martelli GP. Viruses in Artichoke. Advances in Virus Research. 2012;**84**:289-324

[55] Murakami R, Kawano S. A natural host and diversity of *Pepper vein yellows virus* in Japan. Japan Agricultural Research Quarterly. 2017;**51**(1):59-68

[56] Adrian F, Simon L. Pepper vein yellows viruses (PeVYVs). In: Plant Pest Factsheet. UK: Department for Environment Food and Rural Affairs; 2018

[57] Alfaro-Fernandez A, ElShafi EE, Ali MA, El Bashir OOA, Cordoba-Selles MC, Font San Ambrosio MI. First report of *pepper vein yellows virus* infecting hot pepper in Sudan. Plant Disease. 2014;**98**(10):1446

[58] Kim MS, Kim MJ, Hong JS, Choi JK, Ryu KH. Patterns in disease progress and the influence of single and multiple viral infections on pepper (*Capsicum annuum* L.) growth. European Journal of Plant Pathology. 2009;**127**(1):53-61

[59] Arogundade O, Balogun OS, Goodness AU, Kumar PL. Impact of single and double infection with *Cucumber mosaic virus* and *Potato virus Y* on growth and yield of pepper. International Journal of Vegetable Science. 2019;**25**(6):529-541

[60] Alemu T, Hamacher J, Dehne HW. The role of some weeds as hosts of Capsicum viruses in the rift valley parts of Ethiopia. In: 54th International Symposium on Crop Protection, Part I; 7 May, Gent, Belgium. Vol. 67(2). Faculteit Landbouwkundige en Toegepaste Biologische Wetenschappen; 2002. pp. 283-289

[61] Syller J. Facilitative and antagonistic interactions between plant viruses in mixed infections. Molecular Plant Pathology. 2012;**13**:204-216

[62] Arogundade O, Balogun OS, Akinyemi SOS, Kumar PL. Surveys of virus diseases on pepper (*Capsicum* spp.) in South-west Nigeria. African Journal of Biotechnology. 2015;**14**(48):3198-3205

2

Biological Control in *Capsicum* with Microbial Agents

Lorena Barra-Bucarei and Javiera Ortiz

Abstract

Capsicum annuum L. has great importance worldwide for its nutritional characteristics and its antioxidant content. It is cultivated in different geographical areas, under field and greenhouse conditions, and its production can be used for fresh consumption or processing. During its growth, it can be affected by biotic factors, such as pests and diseases that negatively affect the production and quality of its fruits, thus making adequate control measures necessary to avoid relevant economic losses. The environmental conditions that occur in its production promote the development of pests and diseases that can progress rapidly, making it increasingly difficult to manage populations of *Capsicum* . Traditionally, chemical pesticides have been used to deal with these problems, but their indiscriminate use has had negative consequences on the environment and human health. Biological control, based on the use of microorganisms, is thus presented as an efficient and sustainable alternative for *Capsicum* cultivation and offers a series of additional benefits. This chapter reviews the control alternatives available with microbial agents and their applications in the protection of *Capsicum* plants.

Keywords: beneficial microorganisms, entomopathogenic fungi, *Beauveria bassiana*, *Trichoderma*, endophytes

1. Introduction

The genus *Capsicum* belonging to the family *Solanaceae* consists of approximately 31 species, of which only five have been domesticated: *C. annuum*, *C. chinense*, *C. frutescens*, *C. baccatum* and *C. pubescens* [1]. *Capsicum* is known by various names including pepper, chile, chili, chilli, aji, and paprika. Throughout the world, *Capsicum annuum L.* is the most commercially important and widely grown species within this genus. The abundant varieties of *C. annuum*, including sweet peppers and chilli peppers, are important horticultural crops produced worldwide, especially in countries such as Spain and Mexico. Their fruits have remarkable sen-sory attributes in terms of colour, acidity and aroma, as well as an ample diversity of antioxidants, such as phenolic compounds and flavonoids [2]. Moreover, some types present high levels of capsaicin (8-methyl-*n*-vanillyl-6-nonenamide), which provide them with their spicy flavour in addition to therapeutic applications due to their anti-cancer properties [3].

The growth and development of *Capsicum* can be limited by various abiotic and biotic factors that negatively affect its fruit production and quality. It is estimated that losses caused by biotic factors such as invertebrates, pathogens and weeds can vary from 27 to 42%, which would increase to between 48 and 83% if the crops were not protected [4]. Among these, pests and plant diseases seriously affect *Capsicum* crops, in addition to the fact that their production is carried out under limited environmental conditions, which translates into a decrease in yield and quality [5]. For the control of these biotic agents, chemically synthesised pesticides have been traditionally used, which have generated several controversies due to their toxicity in humans and animals, and their damaging effects to the environment. Additionally, they can generate resistance in pathogenic microorganisms [6] and insects [7].

Regarding phytosanitary problems, the high temperatures and high levels of humidity generated in intensive production systems promote the development of fungal diseases that can progress rapidly [8]. Among the diseases that affect *Capsicum*, soil-borne diseases caused by pathogens of the genus *Phytophthora*, *Fusarium*, *Pythium* and *Rhizoctonia* are especially significant. At the fruit level, this crop is affected by pathogens such as *Botrytis cinerea* and *Anthracnose*, the latter caused by a complex of *Colletotrichum* species that is considered a serious problem with heavy losses in fruit yield, exceeding 80% [9]. Another important problem in the cultivation of *Capsicum* is viruses, since there are approximately 70 types that can affect this crop [10], especially the cucumber mosaic virus, pepper mild mottle virus and potato virus Y, among others [11–13]. In general, viruses can interfere with the chlorophyll synthesis of the plant, causing chlorosis and mottling of the foliage (mosaic).

On the other hand, several insect pests affect this crop during the entire growth and production cycle, causing significant yield losses ranging from 50 to 90% [14]. This has resulted in an intensive use of insecticides, mainly chemical. Among the plagues that affect *Capsicum* are: whiteflies, like *Trialeurodes vaporariorum* Westwood (Hemiptera: *Aleyrodidae*) and *Bemisia tabaci* Gennadius (Hemiptera: *Aleyrodidae*); flower thrips, *Frankliniella occidentalis* Pergande (Thysanoptera: *Thripidae*); aphids *Myzus persicae* Sulzer (Hemiptera: *Aphididae*) and *Aphis gossypii* Glover (Hemiptera: *Aphididae*); worms *Helicoverpa armigera* Hübner (*Lepidoptera: Noctuidae*), *Spodoptera litura* Fabricius and *S. exigua* Hübner (Lepidoptera: *Noctuidae*); and mites *Polyphagotarsonemus latus* Banks (Acari: *Tarsonemidae*), *Tetranychus urticae* Koch, *T. ludeni* Zacher and *T. evansi* Baker and Pritchard (Acari: *Tetranychidae*) [15–17]. Within this group, whiteflies, aphids and thrips are con-sidered pests of economic importance worldwide [18]. They have a wide range of hosts, from agricultural species to ornamental plants and are difficult to control due to their high reproductive rate, short life cycle and cryptic behaviour [19–21].

Finally, phytoparasitic nematodes are also organisms present in the cultivation of *Capsicum* and can cause serious problems affecting its performance [22]. Among the nematodes that affect *Capsicum*, those of the genus *Meloidogyne* are especially significant, such as *M. incognita* (Kofoid and White) Chitwood, *M. arenaria* (Neal) Chitwood, *M. enterolobii* (Yang and Eisenback) and *M. javanica* (Treub) Chitwood, which are responsible for root-knot disease [23–25]. These nematodes are found throughout the world, especially in warm areas and in greenhouses. Other nema-todes that also cause economic damage in *Capsicum* include the false root-knot nematode *Nacobbus aberrans* Thorne and Allen [26]. Nematode control strategies are based on the use of chemicals; however, their high cost and principally their toxic effects have led to a search for more sustainable alternatives.

With a growing world population, currently, the main challenge for agriculture is to achieve food security; thus, food production has increased in recent years. However, pests and plant diseases, where *Capsicum* is no exception, have also increased as a result of changing climatic conditions, intensification of production systems and the opening of borders for the free transit of food, including fresh produce. In a conventional way, the high populations of insect pests and plant diseases have been controlled through the use of chemical pesticides. Biological control through the use of antagonistic microorganisms, such as bacteria, fungi and viruses, is presented as an alternative to the use of chemicals. It has received more attention in recent years and arises in response to the search for ways to control pathogens, insects and nematodes in a sustainable manner. However, this type of control must meet some requirements such as being effective against the target organism, not causing problems to the health of people and animals, reaching adequate control levels in the field, the feasibility of being incorporated into integrated management practices and meeting with phytosanitary measures according to the enforced regulations in each country where they are used.

Considering the above, the objective of this chapter is to review the importance of microbial agents in the biological control of pests, plant diseases and nematodes in *Capsicum.*

2. Microbial biological agents

2.1 Biological pest control with microorganisms

In recent years, world markets have expressed increasing concern about the use of agrochemicals. Biological control, which is defined as the reduction of pest populations by natural enemies and usually involves human intervention, is presented as an alternative to the use of chemical pesticides [27]. Biological control agents are classified into predators, parasitoids and pathogens (microorganisms that cause diseases).

After being identified, isolated and reproduced, biocontrol microorganisms are applied in a directed way in dilutions or released on the insect pests and diseases of the crops so that they can carry out their colonising action, produce antagonism, and specific diseases in the agents that require control, with the purpose of reducing the incidence to inoffensive levels [28]. Many microorganisms have been used as biopesticides because they offer a number of additional benefits beyond their target function [29]. An antagonist microorganism used for biological control must meet certain requirements, such as being genetically stable, effective at low concentrations, undemanding in nutrients, adapted to different environmental conditions, effective for a wide range of pathogenic microorganisms, easily grown, easily manipulated, resistant to chemical pesticides, compatible with commercial processes, not phytotoxic and not harmful to humans [30].

Among the microorganisms that have been most studied and reported as antagonists of insect pests and plant pathogens of *Capsicum* are bacteria such as *Bacillus* spp. and *Pseudomonas* spp., and the fungi *Beauveria* spp., *Metarhizium* spp., *Paecilomyces* spp., *Trichoderma* spp. and *Clonostachys* spp., while the viruses of the Baculovirus group have proven to be effective insect controllers. The genus *Bacillus* is made up of species that have been widely used to control insect pests and plant diseases due to the morphological and physiological characteristics that allow them to be ubiquitous in nature. It is especially important to note that species of this genus produce metabolites with antimicrobial properties used for the control of plant pathogens. In addition, the species *B. thuringiensis* Berliner has important qualities for the control of nematodes and protozoa [31, 32]. Another important group used in the control of insect pests in *Capsicum* is the entomopathogenic fungi, which have been widely studied. For more than a century, Pasteur predicted the advantages of entomopathogenic fungi because of their role as bioregulators of pests, acting as parasites to insects that are harmful to plants. Currently, more than 700 species of fungi are known to affect insects of various orders and their use as a biopesticide has increased during the last decade [33]. Mazón [34] indicated that the most important group of entomopathogenic fungi, with practical purposes for pest control, is constituted by *Metarhizium anisopliae* and *Beauveria bassiana.*

2.1.1 Entomopathogenic fungi

Entomopathogenic fungi (EPF) parasitize insects causing serious damage that can even lead to their death. The traditional mode of infecting insects with EPF involves the inoculation of conidia into the cuticle, followed by the formation of a germination tube and an appressorium which, through mechanical and enzymatic action, penetrates the cuticle and reaches the hemocele [35]. After the invasion of the hemocele, these fungi have the ability to re-cross the cuticle of the host and go outside, where they can continue to develop saprophytically on the insect, sporulating and turning them into new foci of dissemination of conidia [36].

EPF have potential for the control of whitefly in *C. annuum* cultivation. The effect of the *B. bassiana* strain M130 on *T. vaporariorum* adults was evaluated under greenhouse conditions reaching control levels of 45.3% in conidia applications to the foliage [14]. Its control of *B. tabaci* was also evaluated; applications of conidia in different concentrations applied to foliage had an effect on the mortality of eggs and nymphs of 37.8–59.04% and 38–75.9%, respectively. The highest percentages were reached with the highest concentrations (2×10^8 conidia/mL) of the Bb01 strain [37]. Flower thrips are an important pest in *Capsicum*; they affect its leaves, flowers and fruits, besides transmitting several viruses, as do white flies. For the control of flower thrips, several EPF have been evaluated as applications of *M. anisopliae* (Met52—Bioglobal Company) by spraying the foliage (flowers and leaves) proved to have the same efficacy on the number of adults of *F. occidentalis* as chemical insecticides [38]. The EPF *Fusarium subglutinans* (strain 12A), applied at concentrations of 1×10^6 conidia/mL, has proven to have the greatest lethal effect (58%) on the second instar nymphs [39]. Other EPF, applied individually or in combinations, have been evaluated in the control of this insect with different levels of efficacy [40]. These fungi have also presented biocontrol action against several aphids. Trials carried out by Curtis et al. [41], where the use of different additives in the formulation of the EPF *Verticillium lecanii* (Zimmerman), now classified as *Lecanicillium lecanii*, was evaluated in applications to pepper foliage for the control of *Myzus persicae*, determined that the formulations did not manage to increase the efficacy of the fungus since the applications of conidia in water suspensions reached the lowest infection rates (<5.0%). The controlling effect of *Lecanicillium attenuatum* (Petch) (CS625) on *Aphis gossypii* has also been evaluated in laboratory trials. The evaluated strain negatively affected aphid populations by decreasing their life expectancy and total fertility, in addition to the direct effect on nymphs [42]. Moreover, these fungi have also proven to be effective in controlling mites. Nugroho and Bin Ibrahim [43] determined that under field conditions, *B. bassiana* was the most effective in suppressing *P. latus* populations (eggs per outbreak) in chilli peppers, in comparison with strains of *M. anisopliae* and *P. fumosoroseus*.

2.1.2 Entomopathogenic nematodes

Entomopathogenic nematodes (EPNs) are widely distributed throughout the world and have a large number of host insects [44]. They are soft-bodied, non-segmented vermiform organisms that are sometimes forced or opt to parasitize insects. They use insects as feeding, dispersal and propagation substrates, characteristics that have allowed them to be used for the biological control of insect pests. Naturally found in the soil, they are able to pursue and locate their host insect, responding to carbon dioxide emissions, vibrations and chemical signals. They have potential for use in crop protection because they are easy to mass produce and harmless to vertebrates and plants [45, 46]. The genera most used in the biological control of insect pests are *Steinernema* spp. and *Heterorhabditis* spp. These nematodes establish a mutualistic association with bacteria of the family Enterobacteriaceae, which are Gram-negative and facultatively anaerobic. Nematodes of the genus *Steinernema* are associated with bacteria of the genus *Xenorhabdus*, while those of the genus *Heterorhabditis* are associated with *Photorhabdus* [47].

Studies by Uhan [48] showed that nematodes of the genus *Steinernema* caused between 23.9 and 78.3% mortality of *S. litura* in pepper plants. The effect of native strains of *Steinernema feltiae* (Filipjev) (Rhabditida: *Steinernematidae*) and *S. carpocapsae* (Weiser) (Rhabditida: *Steinernematidae*) against *S. littoralis* and *Agriotes sordidus* Illiger (Coleoptera: *Elateridae*) larvae and *Ceratitis capitata* (Wiedemann) (Diptera: *Tephritidae*) pupae has also been studied at laboratory and greenhouse levels using pepper plants, with interesting levels of control that varied depending on the strain and soil texture [49]. Although EPNs have been used in *Capsicum* mainly for insects that have larval stages in the soil, they have also been evaluated against pests that spend their entire life cycle on the aerial part of the plant. Studies carried out by Rezaei et al. [50] on pepper plants determined that strains of *S. feltiae* and *H. bacteriophora* (Koppert Biological Systems) had a biocontrol effect on nymphs (second stage) and adults of *T. vaporariorum* in foliar applications.

2.1.3 Entomopathogenic bacteria

One of the most widely used bacteria for insect control is *Bacillus thuringiensis*, a Gram-positive bacterium, which has the ability to form spores as well as synthesise crystalline inclusions containing one or more proteins, some of which are toxic to a significant number of insect pests of the orders Lepidoptera, Diptera, Coleoptera and Hemiptera [51–53]. Evaluations conducted under laboratory conditions determined that 17 of 40 *B. thuringiensis* strains achieved mortality rates of *M. persicae* ranging from 64.4 to 88.9% [54]. On the other hand, some researchers have evaluated the effect of *B. subtilis* strains on this insect; however, no evidence was found of the direct effect of the biocontrol, rather the bacteria was found to promote plant growth by indirectly mitigating the damage caused by the insect [55, 56].

The Gram-negative bacterium *Chromobacterium* subtsugae presents insecticide activity against several orders, highlighting its biocontrol effect on *B. tabaci*, an important pest of *Capsicum*. Its broad-spectrum activity in insect control is related to multiple modes of action that probably involve different chemical compounds [57]. This bacterium could also control populations of *F. occidentalis* [58]. Recently, other bacteria with entomopathogenic action have been studied, such as *Serratia marc-escens*. This bacterium as well as *B. cereus*, *B. pumilus* and some entomopathogenic fungi produce the enzyme chitinase, which has an important action in the pathoge-nicity of insects. Studies carried out by Aggarwal et al. [59] showed that *S. marcescens* (SEN strain) has an important insecticide activity against all stages of development of *S. litura* larvae; the ingestion of this bacterium produced larvae and pupae with less weight, as well as negatively affecting the fertility of the eggs.

Bacteria have also been evaluated for the control of mites. Aksoy et al. [60] investigated the effect of the biocontrol *Pseudomonas putida* (Biotype B) on *T. urticae* at the laboratory level. The bacterium was applied by spraying and dipping newly emerged females, achieving a decrease in the total number of eggs and hatching, compared to their respective controls. Spraying was the most effective method of application. On the other hand, *Burkholderia* spp. (strain A396), after a heat treatment, showed oral toxicity and contact effects against *S. exigua* and *T. urticae*;

its insecticidal action could be related to the production of different bacterial metabolites [61]. There is another group of entomopathogenic bacteria that includes the endosymbionts of the entomopathogenic nematodes, of which the genera *Xenorhabdus* and *Photorhabdus* are especially important [62]. These bacteria can produce metabolites that allow them to colonise and reproduce inside insects.

Although entomopathogenic bacteria have great potential to control insects affecting the cultivation of *C. annuum*, studies with this species are still scarce.

2.2 Disease control with microorganisms

One of the main problems that negatively affect the cultivation of *Capsicum* is the presence of a phytopathogenic complex formed by the genus *Fusarium*, together with others, such as *Phytophthora* spp., *Pythium* spp. and *Rhizoctonia* spp., which are agents that cause seedlings to fall. Together, these can cause losses of between 60 and 100% in production [63]. Nonetheless, conventional control methods have been insufficient and difficult to apply, in addition to the problems associated with soil contamination, phytotoxicity and resistance production in the target pathogen [64]. Considering the above, the use of beneficial microorganisms is presented as an alternative for the control of these pathogens, not only for soil-borne diseases but also for pathogens that can proliferate in the aerial part of the plant and can also com-pensate for the negative environmental impacts caused by chemical pesticides [65]. Studies on the biological control of diseases through the use of microorganisms have recently increased, and the role of the interactions of beneficial microorganisms with the plant and/or the pathogen is receiving more and more attention, not only because of their antagonistic action but also because of their potential to promote plant growth, thus contributing to a more sustainable production over time [66].

Among the most commonly used microorganisms for the control of pathogens are fungi of the genera *Trichoderma*, *Clonostachys*, and *Penicillium*; and the bacteria *Pseudomonas* and *Bacillus*. In relation to their antagonistic activity, these present diverse mechanisms of action that can act independently or jointly. If the antagonist can express different mechanisms of action, the likelihood of developing resistance in the pathogen is reduced. The mechanisms of action described include competi-tion for nutrients and space, antibiosis, parasitism (considered as an antagonistic symbiosis between organisms) and activation of systemic resistance. The use of these control agents is projected as a more efficient control alternative; however, there are several aspects that should be studied so that they can express their full potential, such as the form of application, the combined use of antagonists, formulation and conditions during application [67]. Moreover, the possibility that beneficial microorganisms may persist in the soil could be an important advantage over the use of pesticides. An additional advantage is that if they remain in the rhizosphere they are in the first line of defence against attack by soil pathogens.

2.2.1 Antagonistic fungi

A diverse group of fungi have been shown to antagonise a significant number of pathogens affecting *Capsicum,* such as *Phytophthora* spp., *Colletotrichum* spp., *Rhizoctonia* spp., *Fusarium* spp., and *Botrytis cinerea*, among others [68–72]. One of the most cited examples in the literature are the fungi of the genus *Trichoderma*, which due to their diversity, versatility, adaptability and easy handling have proven to be good candidates for commercial development as a biofungicide. Moreover, these fungi produce three types of propagules, hyphae, chlamydospores and conidia, all of which are active against pathogens in different phases of their life cycle, along with presenting an interesting activity promoting plant growth.

Studies carried out by Ahmed [68] indicate that *T. harzianum* presents an antagonistic effect against the pathogen *P. capsici* because it significantly reduces stem necrosis compared to values obtained in plants inoculated only with the pathogen. Applications of *Trichoderma* spp. in jalapeño peppers were able to decrease the growth rate of *P. capsici* in addition to increasing plant growth [73]. Another soil fungus controlled by *Trichoderma* is *Fusarium* spp., which was confirmed in studies carried out by Sinha et al. [74] on chilli peppers under greenhouse conditions, which showed that strains of *T. viride* and *T. harzianum* reduced the incidence of *F. oxysporum* f. sp. *capsici* by 83.93–87.5%. *Trichoderma* has also been used to control anthracnose, which is caused by several species of *Colletotrichum*. Applications of strains of *T. harzianum* (BHUF4) and *T. asperellum* (T16A) promote the accumulation of the phenol that stimulates the expression of the defence gene, both of which have proven to induce and acquire systemic resistance, providing a solid protection against *C. truncatum* [72]. Studies carried out by Rahman et al. [71] also determined that *Trichoderma* spp. had a biocontrol effect on *C. capsici* in chilli peppers since they managed to decrease the percentage of infection in fruit, although the strain *T. harzianum* (IMI-392433) presented the best antagonistic effect against this pathogen, all the strains (*T. harzianum, T. viride* and *T. pseudokoningii*) presented a promoter effect in plant growth and higher fruit yield.

The *Clonostachys* fungus also presents important characteristics in the biological control of pathogenic fungi. Nobre et al. [69] determined that strains of *C. rosea* were able to suppress more than 80% of the sporulation of *B. cinerea* in chilli peppers and tomatoes. Another fungus studied for disease control in *Capsicum* is *Xylaria poitei*. The antagonistic activity *in vivo* and the protection that *Xylaria poitei* provided chilli pepper seedlings against the pathogen *P. capsici* were evaluated. This antagonist allowed the survival of 58.3% of the seedlings, while seedlings inoculated only with the pathogen showed 100% mortality [75].

2.2.2 Antagonistic bacteria

Several species of *Bacillus* are known for their antagonistic capacity against pathogens that cause disease in plants; many of these are also considered because they could have a direct action in disease control as a result of the production of metabolites or indirectly by promoting growth. These bacteria act not only on soil-borne diseases but also on leaf and fruit pathogens [76–78]. Diverse studies have provided evidence that plant growth-promoting bacteria have the ability to activate systemic resistance induced by the activation of physical or chemical defences of the host plant by an inducing agent [79].

Studies carried out by Yu et al. [77] determined that applications of *B. subtilis* (CAS15 strain) significantly suppressed spore germination of *F. oxysporum* f. sp. *capsici* by 8–64%. In addition, the incidence of the disease was much lower in plants treated with the bacteria compared to the control (12.5–56.9%); these results are attributed to the induction of systemic resistance. This bacterium has also been used in *Capsicum* for the control of the anthracnose disease in chilli plants caused by *Colletotrichum gloeosporioides* (OGC1 strain). Seeds treated with the bacterium showed a 65% decrease in disease incidence compared to seeds treated with the pathogen alone [78]. Some *Bacillus* species have also been evaluated in the control of diseases caused by pathogenic bacteria, such as pepper bacterial wilt caused by *Ralstonia solanacearum*, an important *Capsicum* disease [79]. The application of *B. amyloliquefa-ciens* (strain Bg-C31) and its metabolites significantly reduced the percentage of plants affected by *R. solanacearum* at the field level and in pepper pots [80].

Pseudomonas is another group of bacteria used for disease control in *Capsicum*; among these, the most reported is *P. fluorescens* which has been shown to have a biocontrol effect on pathogens and also some phytoparasitic nematodes [81, 82]. Basu [83] evaluated the effect of *P. fluorescens* on *Ralstonia solanacearum* in chilli pepper plants and determined that combined applications of the antago-nist to the seed and the soil resulted in the lowest percentages of disease inci-dence (0.83–10.82%), compared to the control which was inoculated only with the pathogen (9.99–44.96%).

Many bacteria have demonstrated an important potential for the control of diseases caused by fungi and bacteria; nevertheless, many of them have been evaluated under controlled conditions, which makes it necessary to concentrate efforts to increase the number of field level trials, and thus determine the real potential of being incorporated into the cultivation of *Capsicum.*

2.3 Nematodes with microorganisms

Infestation of peppers by nematodes, such as *Meloidogyne* spp., is one of the major problems for pepper production worldwide [84]. They attack the roots, leading to root system dysfunction, reduced rooting volume, inefficient use of water and nutrients, reduced crop and plant growth and yield [85]. Another nematode that causes damage in *Capsicum* is the false root-knot nematode, *Nacobbus aberrans*, which is associated with *F. oxysporum* and *P. capsici*, and can cause up to 100% crop loss [86]. For its control, chemical nematicides are applied; however, in the cultivation of *C. annuum*, its use has been restricted due to its toxicity levels, residual elements in the environment and the selection of resistant nematode populations [87].

An alternative for the biological control of phytoparasitic nematodes is nematophagous fungi. These ubiquitous microorganisms are capable of modifying their saprophytic behaviour and can feed on nematodes in unfavourable nutritional conditions. They are natural enemies of nematodes and have developed highly sophisticated infection strategies [88]. They are classified into three groups according to their predation characteristics. First are the nematode traps or predators (their modified hyphae form traps and through a chemical and mechanical process they digest the nematode), second are the opportunists or ovicides (they form traps to catch eggs, cysts and females) and finally the endoparasites (forced parasites of the nematodes that use their spores as structures of infection, which can adhere to the cuticle of the nematode or be ingested) [89, 90].

Paecilomyces lilacinus have a negative effect on the root nodulation produced by *M. incognita* in peppers [91]. Another trial used *B. bassiana* strains and determined that applications of this fungus to the substrate in pepper plants decrease the number of root nodules caused by *M. incognita* [92]. The combination of nematophagous fungi has also been evaluated on *M. incognita*. Studies carried out by Requena Candela [93] demonstrated the antagonistic effect of the combination of *P. lilacinus* and *T. harzianum* by reaching decreases in the incidence of nematodes close to 70%. Pérez-Rodríguez et al. [94] evaluated the nematophagous action of the fungus *Pochonia chlamydosporia* for the control of *N. aberrans* nematodes in broad chilli peppers and determined that the combination of vermicompost with the nematophagous achieves a reduction in the number of juvenile nematodes, the number of eggs and females per gram of roots, as well as increasing the dry matter of the plants.

2.4 Endophytic microorganisms for the control of insect pests and plant diseases

Endophytes are defined as microorganisms that spend most or all of their life cycle colonising host plant tissues without causing obvious damage [95].

Endophytes are associated with most plant species, are naturally found in the ecosystem and are considered an extremely important partner for plants [96, 97]. These microorganisms have been of interest for study during the last few years due to the beneficial characteristics they are able to bestow onto their hosts [98], among which are the promotion of plant growth, inhibition of pathogenic organisms, control of insect pests, removal of contaminants from the soil and increased tolerance to extreme conditions of temperature, water availability and salinity [99–101]. Consequently, they have an important future in agrifood production.

Currently, several studies have shown that endophytic fungi can protect host plants against pathogens and herbivores [102, 103]. Some endophytes, when artificially inoculated, can confer the beneficial characteristics mentioned above to their hosts, as they can influence key aspects of physiology. The host plant receives multiple benefits from its interaction with the endophyte in exchange for carbon-based resources [104]. The type of interaction used by endophyte biological control-lers is not yet clear [105], but it is believed that this type of interaction is established through a process of co-evolution [106].

These fungi generate interspecific interactions and protection against the inci-dence of pathogens produced by direct mechanisms such as competition, parasitism, antibiosis and indirect mechanisms such as induction of resistance [97, 104]. The activation of systemic resistance in the plant may be due to the presence of endophytic fungi, a mechanism that has already been shown to be effective against other fungal pathogens [107]. On the other hand, among the most important reported mecha-nisms for pest control are parasitism by ingestion [108], antagonism by the action of metabolites [109, 110], systemic resistance [111] and a tritrophic action [112].

Works carried out by Barra-Bucarei et al. [113] present the first report of the endophyte *B. bassiana* with antagonistic action against *Botrytis cinerea* in chilli pepper plants. Five native strains were evaluated, and the percentage of leaf area affected by the pathogen (PSAP) was determined. Plants inoculated with endophytes showed 2 to 18% of PSAP compared with plants treated only with the pathogen, which exhibited early symptoms of the disease with a PSAP of 63%. Another study conducted by the same authors (unpublished data) demonstrated that the same strains are endophytic (**Figure 1**) and reduced the symptoms caused

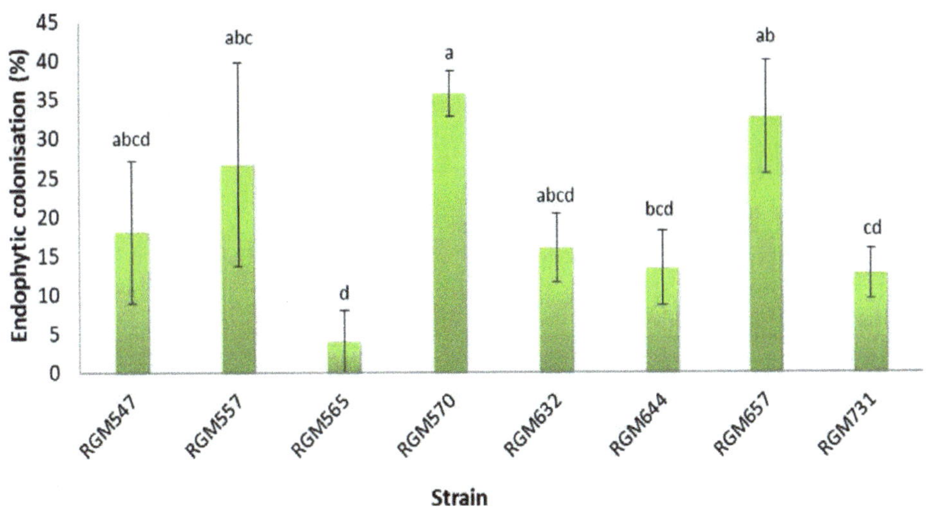

Figure 1.
Endophytic colonisation of B. bassiana *strains in chilli pepper plants (leaves, stems and roots), 30 days after inoculation (n = 5). Data represent the mean standard error. Different letters over the bars represent significant differences among the treatments according to the Fisher's LSD test (p < 0.05).*

Figure2.
In vivo *antagonism of* B. bassiana *(565) against* F. oxysporum *(Fo). Plants inoculated with* F. oxysporum *(Fo) showed chlorosis and plants inoculated with* B. bassiana *RGM565 +* Fo *showed no symptoms, 45 days after inoculation (n = 10).*

by *F. oxysporum* in chilli pepper plants, increasing dry matter and decreasing the level of leaf chlorosis in adult plants, making the plants more resistant to the attack of this pathogen (**Figure 2**). These studies coincide with research conducted by Jaber and Alananbeh [103], in pepper plants which provided evidence that commercial strains of *B. bassiana* (Naturalis) and *M. brunneum* (Bipesco5) can inhibit the growth of several species of *Fusarium* (*F. oxysporum*, *F. culmorum* and *F. moniliforme*) *in vitro* and in potting trials.

3. Conclusion

With the increase of the world population, it is urgent for the development of sustainable strategies to improve food availability. *Capsicum annuum* is an excellent source of natural health-related compounds, such as micronutrients and antioxidants; its fruits have been used for fresh and cooked consumption. The crop of this vegetable can be negatively affected by biotic factors, provoking decreased yields. In this chapter reviewed, we have provided evidence of the potential that microorganisms present for the control of insect pests, plant diseases, and nematodes in the *Capsicum* crop. There are a significant number of investigations carried out under laboratory conditions with good biocontrol results; however, it is necessary to increase the number of field studies and application methods, since environmental conditions could condition the behaviour of microorganisms and not reach the same levels of efficacy obtained under controlled conditions. In addition, it is necessary to study the ecological behaviour of beneficial microorganisms and their interaction with others. The biocontrol microbial agents can fulfil diverse functions in plants and give an eco-friendly approach to promoting sustainable agriculture.

Author details

Lorena Barra-Bucarei[1*] and Javiera Ortiz[2]

1 Instituto de Investigaciones Agropecuarias, Chillán, Chile

2 Universidad de Concepción, Chillán, Chile

*Address all correspondence to: lorena.barra@inia.cl

References

[1] Mongkolporn O, Taylor PW. *Capsicum*. In: Wild Crop Relatives: Genomic and Breeding Resources. Berlin, Heidelberg: Springer; 2011. pp. 43-57

[2] Estrada B, Bernal MA, Díaz J, Pomar F, Merino F. Capsaicinoids in vegetative organs of *Capsicum annuum* L. in relation to fruiting. Journal of Agricultural and Food Chemistry. 2002;**50**(5):1188-1191

[3] Surh YJ, Lee SS. Capsaicin in hot chili pepper: Carcinogen, co-carcinogen or anticarcinogen? Food and Chemical Toxicology. 1996;**34**(3):313-316

[4] Oerke EC, Dehne HW. Safeguarding production—Losses in major crops and the role of crop protection. Crop Protection. 2004;**23**(4):275-285

[5] López-Marín J, González A, Pérez-Alfocea F, Egea-Gilabert C, Fernández JA. Grafting is an efficient alternative to shading screens to alleviate thermal stress in greenhouse-grown sweet pepper. Scientia Horticulturae. 2013;**149**:39-46

[6] Zhu J, Wang J, Ding Y, Liu B, Xiao W. A systems-level approach for investigating organophosphorus pesticide toxicity. Ecotoxicology and Environmental Safety. 2018;**49**:26-35

[7] Jensen SE. Insecticide resistance in the western flower thrips, *Frankliniella occidentalis*. Integrated Pest Management Reviews. 2000;**5**(2):131-146

[8] Punja ZK, Rodriguez G, Tirajoh A, Formby S. Role of fruit surface mycoflora, wounding and storage conditions on post-harvest disease development on greenhouse tomatoes. Canadian Journal of Plant Pathology. 2016;**38**(4):448-459

[9] Mahasuk P, Taylor PWJ, Mongkolporn O. Identification of two new genes conferring resistance to *Colletotrichum acutatum* in *Capsicum baccatum*. Phytopathology. 2009;**99**(9):1100-1104

[10] Pernezny K, Elliott M, Palmateer A, Havranek N. Guidelines for identification and management of plant disease problems: Part II. Diagnosing plant diseases caused by fungi, bacteria and viruses [Internet]. Florida: University of Florida, Plant Pathology Department, UF/IFAS Extension publication; 2016. pp. 1-8. Available from: http://edis.ifas.ufl.edu

[11] Chaim AB, Grube RC, Lapidot M, Jahn M, Paran I. Identification of quantitative trait loci associated with resistance to cucumber mosaic virus in *Capsicum annuum*. Theoretical and Applied Genetics. 2001;**102**(8):1213-1220

[12] Genda Y, Sato K, Nunomura O, Hirabayashi T, Ohnishi J, Tsuda S. Immunolocalization of pepper mild mottle virus in *Capsicum annuum* seeds. Journal of General Plant Pathology. 2005;**71**(3):238-242

[13] Moodley V, Ibaba JD, Naidoo R, Gubba A. Full-genome analyses of a potato virus Y (PVY) isolate infecting pepper (*Capsicum annuum* L.) in the Republic of South Africa. Virus Genes. 2014;**49**(3):466-476

[14] Kumar NKK. Yield loss in chilli and sweet pepper due to *Scirtothrips dorsalis* Hood (Thysanoptera: *Thripidae*). Pest Management in Horticulture Ecosystem. 1995;**1**(2):61-69

[15] Reddy DNR. Pests infesting chilli (*Capsicum annuum* L.) - II. In the transplanted crop. Mysore Journal of Agricultural Sciences. 1983;**17**(3):246-251

[16] Venzon M, Amaral DSSL, Perez AL, Cruz FAR, Togni PHB. Oliveira RM Identificação e manejo ecológico de pragas da cultura da pimenta. Belo Horizonte: EPAMIG; 2011. p. 40

[17] Meena RS, Ameta OP, Meena BL. Population dynamics of sucking pests and their correlation with weather parameters in chilli, *Capsicum annuum* L. crop. The Bioscan. 2013;**8**(1):177-180

[18] Hodges GS, Evans GA. An identification guide to the whiteflies (Hemiptera: *Aleyrodidae*) of the Southeastern United States. Florida Entomologist. 2005:**88**(4);518-534

[19] Choi WI, Lee EH, Choi BR, Park HM, Ahn YJ. Toxicity of plant essential oils to *Trialeurodes vaporariorum* (Homoptera: *Aleyrodidae*). Journal of Economic Entomology. 2003;**96**(5):1479-1484

[20] Cloyd RA. Western flower thrips (*Frankliniella occidentalis*) management on ornamental crops grown in greenhouses: Have we reached an impasse. Pest Technology. 2009;**3**(1):1-9

[21] Kim CS, Lee JB, Kim BS, Nam YH, Shin KS, Kim JW, et al. Una técnica para la prevención de la mosca blanca de invernadero (*Trialeurodes vaporariorum*) utilizando el hongo entomopatógeno *Beauveria bassiana* M130. Journal of Microbiology and Biotechnology. 2014;**24**(1):1-7

[22] Shahid S, Khan A. Spatial pattern analysis of three nematod populations associated with chilli. Fundamental and Applied Nematology. 1993;**16**(5):473-478

[23] Khan Z, Tiyagi SA, Mahmood I, Rizvi R. Effects of N fertilisation, organic matter, and biofertilisers on the growth and yield of chilli in relation to management of plant-parasitic nematodes. Turkish Journal of Botany. 2012;**36**(1):73-81

[24] Khan B, Khan AA, Khan MR. Pathogenic variability among isolates of *Meloidogyne javanica* on *Capsicum annuum*. Journal of Nematology. 2003;**35**(4):430

[25] Villar-Luna E, Goméz-Rodriguez O, Rojas-Martínez RI, Zavaleta-Mejía E. Presence of *Meloidogyne enterolobii* on Jalapeño pepper (*Capsicum annuum* L.) in Sinaloa, Mexico. Helminthologia. 2016;**53**(2):155-160

[26] Djian-Caporalino C. Root-knot nematodes (*Meloidogyne* spp.), a growing problem in French vegetable crops. EPPO Bulletin. 2012;**42**(1):127-137

[27] Hoffmann MP, Frodsham A. Natural Enemies of Vegetable Insect Pests. Ithaca: Cooperative Extension, Cornell University; 1993

[28] Suquilanda M. La Producción Orgánica de cultivos en el Ecuador. In: Presentado en el Seminario "Procesos de certificación y producción orgánica". Ecuador: AGEAPSE; 2001

[29] Glare T, Caradus J, Gelernter W, Jackson T, Keyhani N, Köhl J, et al. Have biopesticides come of age? Trends in Biotechnology. 2012;**30**(5):250-258

[30] Spadaro D, Gullino ML. State of the art and future prospects of the biological control of postharvest fruit diseases. International Journal of Food Microbiology. 2004;**91**(2):185-194

[31] Vankova J. Study of the effect of *Bacillus thuringiensis* on insects. Folia Biologica. 1957;**3**:175-182

[32] Smith RA, Couche GA. The phylloplane as a source of *Bacillus thuringiensis* variants. Applied and Environmental Microbiology. 1991;**57**(1):311-315

[33] Shah FA, Ansari MA, Watkins J, Phelps Z, Cross J, Butt TM. Influence

of commercial fungicides on the germination, growth and virulence of four species of entomopathogenic fungi. Biocontrol Science and Technology. 2009;**19**(7):743-753

[34] Mazón A. Producción, uso y control de calidad de hongos entomopatógenos en Nicaragua. In: Manejo Integrado de plagas. Vol. 63. Costa Rica: CATIE; 2001. pp. 95-103

[35] Hajek AE, St. Leger RJ. Interactions between fungal pathogens and insect hosts. Annual Review of Entomology. 1994;**39**(1):293-322

[36] Meyling NV, Eilenberg J. Ecology of the entomopathogenic fungi *Beauveria bassiana* and *Metarhizium anisopliae* in temperate agroecosystems: Potential for conservation biological control. Biological Control. 2007;**43**(2):145-155

[37] Zafar J, Freed S, Khan BA, Farooq M. Effectiveness of *Beauveria bassiana* against cotton whitefly, *Bemisia tabaci* (Gennadius) (*Aleyrodidae*: Homoptera) on different host plants. Pakistan Journal of Zoology. 2016;**48**(1):91-99

[38] Dura O, Dura S, Bilen E. Determination of the effective dose of *Metarhizium anisopliae* str. F 52 for biological control of western flower thrips on pepper plant. Bahçe. 2012;**41**(1):1-7

[39] Demirözer O, Uzun A, Arici ŞE, Gep İ, Bakay R. Insecticidal effect of *Fusarium subglutinans* on *Frankliniella occidentalis* (Pergande) (Thysanoptera: *Thripidae*). Hellenic Plant Protection Journal. 2016;**9**(2):66-72

[40] Gao Y, Reitz SR, Wang J, Xu X, Lei Z. Potential of a strain of the entomopathogenic fungus *Beauveria bassiana* (Hypocreales: *Cordycipitaceae*) as a biological control agent against western flower thrips, *Frankliniella occidentalis* (Thysanoptera: *Thripidae*). Biocontrol Science and Technology. 2012;**22**(4):491-495

[41] Curtis JE, Price TV, Ridland PM. Initial development of a spray formulation which promotes germination and growth of the fungal entomopathogen *Verticillium lecanii* (Deuteromycotina: Hyphomycetes) on *Capsicum* leaves (*Capsicum annuum* grossum) and infection of *Myzus persicae* Sulzer (Homoptera: *Aphididae*). Biocontrol Science and Technology. 2003;**13**(1):47-63

[42] Kim JJ. Influence of *Lecanicillium attenuatum* on the development and reproduction of the cotton aphid, *Aphis gossypii*. BioControl. 2007;**52**(6):789-799

[43] Nugroho I, Bin Ibrahim Y. Efficacy of laboratory prepared wettable powder formulation of entomopathogenous fungi *Beauveria bassiana*, *Metarhizium anisopliae* and *Paecilomyces fumosoroseus* against the *Polyphagotarsonemus latus* (Bank) (Acari: Tarsonemidae) (broad mite) on *Capsicum annuum* (chilli). Journal of Biosciences. 2007;**18**(1):1-11

[44] Hominick WM. Biogeography. Entomopathogenic Nematology. 2002;**1**:115-143

[45] Grewal PS. Formulation and Application Technology. In: Entomopathogenic Nematology. CABI Publishing; 2002. pp. 265-287

[46] Dillman AR, Sternberg PW. Entomopathogenic nematodes. Current Biology. 2012;**22**(11):R430-R431

[47] Godjo A, Afouda L, Baimey H, Decraemer W, Willems A. Molecular diversity of *Photorhabdus* and *Xenorhabdus* bacteria, symbionts of *Heterorhabditis* and *Steinernema* nematodes retrieved from soil in Benin. Archives of Microbiology. 2018;**200**(4):589-601

[48] Uhan TS. Bioefikasi beberapa isolat nematoda entomopatogenik *Steinernema*

spp. terhadap *Spodoptera litura* Fabricius pada tanaman cabai di rumah kaca. Jurnal Hortikultura. 2008;**18**(2):175-184

[49] Campos-Herrera R, Gutiérrez C. Screening Spanish isolates of steinernematid nematodes for use as biological control agents through laboratory and greenhouse microcosm studies. Journal of Invertebrate Pathology. 2009;**100**(2):100-105

[50] Rezaei N, Karimi J, Hosseini M, Goldani M, Campos-Herrera R. Pathogenicity of two species of entomopathogenic nematodes against the greenhouse whitefly, *Trialeurodes vaporariorum* (Hemiptera: *Aleyrodidae*), in laboratory and greenhouse experiments. Journal of Nematology. 2015;**47**(1):60

[51] MacIntosh SC, Stone TB, Sims SR, Hunst PL, Greenplate JT, Marrone PG, et al. Specificity and efficacy of purified *Bacillus thuringiensis* proteins against agronomically important insects. Journal of Invertebrate Pathology. 1990;**56**(2):258-266

[52] Porcar M, Grenier AM, Federici B, Rahbé Y. Effects of *Bacillus thuringiensis* δ-endotoxins on the pea aphid (*Acyrthosiphon pisum*). Applied and Environmental Microbiology. 2009;**75**(14):4897-4900

[53] Palma L, Muñoz D, Berry C, Murillo J, De Escudero IR, Caballero P. Molecular and insecticidal characterization of a novel Cry-related protein from *Bacillus thuringiensis* toxic against *Myzus persicae*. Toxins. 2014;**6**(11):3144-3156

[54] Torres-Quintero MC, Arenas- Sosa I, Hernández-Velázquez VM, Suárez-Rodríguez R, Peña-Chora G. Characterization of *Bacillus thuringiensis* (*Bacillaceae*) strains pathogenic to *Myzus persicae* (Hemiptera: *Aphididae*). Florida Entomologist. 2016;**99**(4):639-643

[55] Herman MAB, Nault BA, Smart CD. Effects of plant growth-promoting rhizobacteria on bell pepper production and green peach aphid infestations in New York. Crop Protection. 2008;**27**(6):996-1002

[56] Mardani-Talaee M, Razmjou J, Nouri-Ganbalani G, Hassanpour M, Naseri B. Impact of chemical, organic and bio-fertilizers application on bell pepper, *Capsicum annuum* L. and biological parameters of *Myzus persicae* (Sulzer) (Hem.: Aphididae). Neotropical Entomology. 2017;**46**(5):578-586

[57] Ruiu L. Insect pathogenic bacteria in integrated pest management. Insects. 2015;**6**(2):352-367

[58] Kivett JM. Efficacy of entomopathogenic organisms *Beauveria bassiana*, *Isaria fumosoroseus*, *Metarhizium anisopliae* and *Chromobacterium subtsugae* against the western flower thrips, *Frankliniella occidentalis*, under both laboratory and greenhouse conditions [doctoral dissertation]. Kansas: Kansas State University; 2015

[59] Aggarwal C, Paul S, Tripathi V, Paul B, Khan MA. Chitinolytic activity in *Serratia marcescens* (strain SEN) and potency against different larval instars of *Spodoptera litura* with effect of sublethal doses on insect development. BioControl. 2015;**60**(5):631-640

[60] Aksoy HM, Ozman-Sullivan SK, Ocal H, Celik N, Sullivan GT. The effects of *Pseudomonas putida* biotype B on *Tetranychus urticae* (Acari: *Tetranychidae*). In: Diseases of Mites and Ticks. Dordrecht: Springer; 2008. pp. 223-230

[61] Cordova-Kreylos AL, Fernandez LE, Koivunen M, Yang A, Flor-Weiler L, Marrone PG. Isolation and characterization of *Burkholderia rinojensis* sp. nov., a non- *Burkholderia*

cepacia complex soil bacterium with insecticidal and miticidal activities. Applied and Environmental Microbiology. 2013;**79**(24):7669-7678

[62] Burnell A, Stock SP. *Heterorhabditis*, *Steinernema* and their bacterial symbionts—Lethal pathogens of insects. Nematology. 2000;**2**(1):31-42

[63] Villa-Martínez A, Pérez-Leal R, Morales-Morales HA, Basurto-Sotelo M, Soto-Parra JM, Martínez-Escudero E. Situación actual en el control de *Fusarium* spp. y evaluación de la actividad antifúngica de extractos vegetales. Acta Agronómica. 2015;**64**(2):194-205

[64] Rajkumar M, Lee WH, Lee KJ. Screening of bacterial antagonists for biological control of *Phytophthora* blight of pepper. Journal of Basic Microbiology: An International Journal on Biochemistry, Physiology, Genetics, Morphology, and Ecology of Microorganisms. 2005;**45**(1):55-63

[65] Das MM, Haridas M, Sabu A. Biological control of black pepper and ginger pathogens, *Fusarium oxysporum*, *Rhizoctonia solani* and *Phytophthora capsici*, using *Trichoderma* spp. Biocatalysis and Agricultural Biotechnology. 2019;**17**:177-183

[66] Gautam AK, Shubhi A. Fungal endophytes: Potential biocontrol agents in agriculture. In: Kumar A, Singh AK, Choudrary KK, editors. Role of Plant Growth Promoting Microorganisms in Sustainable Agriculture and Nanotechnology. Duxford, UK: Woodhead Publishing; 2019. pp. 241-283

[67] Abeysinghe S. Effect of combined use of *Bacillus subtilis* CA32 and *Trichoderma harzianum* RU01 on biological control of *Rhizoctonia solani* on *Solanum melongena* and *Capsicum annuum*. Plant Pathology Journal (Faisalabad). 2009;**8**(1):9-16

[68] Ahmed AS, Ezziyyani M, Sánchez CP, Candela ME. Effect of chitin on biological control activity of *Bacillus* spp. and *Trichoderma harzianum* against root rot disease in pepper (*Capsicum annuum*) plants. European Journal of Plant Pathology. 2003;**109**(6):633-637

[69] Nobre SA, Maffia LA, Mizubuti ES, Cota LV, Dias APS. Selection of *Clonostachys rosea* isolates from Brazilian ecosystems effective in controlling *Botrytis cinerea*. Biological Control. 2005;**34**(2):132-143

[70] Pandey AP, Zacharia S, Patil NB. Comparative efficacy of different methods of application of *Trichoderma harazianum* Rifai. in the management of dry root rot of chilli (*Capsicum annuum*) caused by *Rhizoctonia solani* (kuhn). International Journal of Current Microbiology and Applied Sciences. 2017;**6**(3):482-489

[71] Rahman MA, Ansari TH, Alam MF, Moni JR, Ahmed M. Efficacy of *Trichoderma* against *Colletotrichum capsici* causing fruit rot due to anthracnose of chili (*Capsicum annuum* L.). The Agriculturists. 2018;**16**(02):75-87

[72] Saxena A, Mishra S, Ray S, Raghuwanshi R, Singh HB. Differential reprogramming of defense network in *Capsicum annuum* L. plants against *Colletotrichum truncatum* infection by phyllospheric and rhizospheric *Trichoderma* strains. Journal of Plant Growth Regulation. 2020;**39**:751-763

[73] López CG, González PAG. Selección de cepas nativas de *Trichoderma* spp. con actividad antagónica sobre *Phytophthora capsici* Leonian y promotoras de crecimiento en el cultivo de chile (*Capsicum annuum* L.). Revista Mexicana de Fitopatología. 2004;**22**(1):117-124

[74] Sinha A, Harshita D, Singh R, Verma A. Bioefficacy of *Trichoderma*

harzianum and *Trichoderma viride* against *Fusarium oxysporum* f. sp. *capsici* causing wilt disease in chilli. Journal of Pharmacognosy and Phytochemistry. 2018;**7**(5):965-966

[75] Ramos-Sandoval RU, Gutiérrez-Soto JG, Rodríguez- Guerra R, Salcedo-Martínez SM, Hernández-Luna CE, Luna-Olvera HA, et al. Antagonismo de dos ascomicetos contra *Phytophthora capsici* Leonian, causante de la marchitez del chile (*Capsicum annuum* L.). Revista mexicana de fitopatología. 2010;**28**(2):75-86

[76] El-Hamshary OIM, Khattab AA. Evaluation of antimicrobial activity of *Bacillus subtilis* and *Bacillus cereus* and their fusants against *Fusarium solani*. Research Journal of Cell and Molecular Biology. 2008;**2**:24-29

[77] Yu X, Ai C, Xin L, Zhou G. The siderophore-producing bacterium, *Bacillus subtilis* CAS15, has a biocontrol effect on *Fusarium* wilt and promotes the growth of pepper. European Journal of Soil Biology. 2011;**47**(2):138-145

[78] Ashwini N, Srividya S. Potentiality of *Bacillus subtilis* as biocontrol agent for management of anthracnose disease of chilli caused by *Colletotrichum gloeosporioides* OGC1. 3 Biotech. 2014;**4**(2):127-136

[79] Knapp S, Bohs L, Nee M, Spooner DM. *Solanaceae*—A model for linking genomics with biodiversity. Comparative and Functional Genomics. 2004;**5**(3):285-291

[80] Hu HQ , Li XS, He H. Characterization of an antimicrobial material from a newly isolated *Bacillus amyloliquefaciens* from mangrove for biocontrol of *Capsicum* bacterial wilt. Biological Control. 2010;**54**(3):359-365

[81] Abuzar S. Antagonistic effects of some fluorescent *Pseudomonas* strains against root rot fungi (*Rhizoctonia solani* and *Fusarium oxysporum*) and root-knot nematodes (*Meloidogyne incognita*) on chili (*Capsicum annuum*). World Applied Sciences Journal. 2013;**27**(11):1455-1460

[82] Thiyagarajan SS, Kuppusamy H. Biological control of root knot nematodes in chillies through *Pseudomonas fluorescens*'s antagonistic mechanism. Plant Science. 2014;**2**:152-158

[83] Basu A. Bio-efficacy of *Pseudomonas fluorescens* (7% WP and 5% SC formulations) against bacterial wilt disease of chili. Asia Pacific Journal of Sustainable Agriculture, Food and Energy. 2014;**2**(2):36-40

[84] Affokpon A, Coyne DL, Htay CC, Agbèdè RD, Lawouin L, Coosemans J. Biocontrol potential of native *Trichoderma* isolates against root-knot nematodes in West African vegetable production systems. Soil Biology and Biochemistry. 2011;**43**(3):600-608

[85] Kong C, Zhai C, Li J, Zhang J, Zhang K, Liu Y. Evaluation of *Stropharia* sp. 1.2052 nematicidal effects against *Meloidogyne incognita* on tomato. African Journal of Microbiology Research. 2013;7(50):5737-5741

[86] Lax P, Dueñas JCR, Ramos D, Doucet ME, Braga R, Kobori R. Host suitability of peppers to the false root-knot nematode *Nacobbus aberrans*. Crop Protection. 2016;**79**:15-19

[87] Xie H, Yan D, Mao L, Wang Q, Li Y, Ouyang C, et al. Evaluation of methyl bromide alternatives efficacy against soil-borne pathogens, nematodes and soil microbial community. PLoS One. 2015;**10**(2):e0117980

[88] De Freitas Soares FE, Sufiate BL, de Queiroz JH. Nematophagous fungi: Far beyond the endoparasite, predator

and ovicidal groups. Agriculture and Natural Resources. 2018;**52**(1):1-8

[89] Braga FR, de Araújo JV. Nematophagous fungi for biological control of gastrointestinal nematodes in domestic animals. Applied Microbiology and Biotechnology. 2014;**98**(1):71-82

[90] Hamza MA, Lakhtar H, Tazi H, Moukhli A, Fossati-Gaschignard O, Miche L, et al. Diversity of nematophagous fungi in Moroccan olive nurseries: Highlighting prey-predator interactions and efficient strains against root-knot nematodes. Biological Control. 2017;**114**:14-23

[91] Santillán GF, Rebaza LC, Ruiz JC. Eficacia de *Paecilomyces lilacinus* en el control de *Meloidogyne incognita* que ataca al cultivo de *Capsicum annuum*, "pimiento piquillo". Fitosanidad. 2016;**20**(3):109-119

[92] Fernández Bocanegra FJ. Efecto de *Beauveria bassiana* sobre *Meloidogyne incognita* en el número de nódulos radiculares en plantas de *Capsicum annuum* L. cv. Piquillo cultivadas en invernadero [Tesis Biólogo-microbiólogo]. Trujillo, Perú: Universidad Nacional de Trujillo; 2015. Available from: http://dspace.unitru.edu.pe/handle/UNITRU/4545

[93] Requena Candela AM. Control biológico de *Meloidogyne incognita* en pimiento (*Capsicum annuum*) [Tesis doctoral]. Cartagena, España: Universidad Politécnica de Cartagena; 2013. Available from: https://repositorio. upct.es/handle/10317/4019

[94] Perez-Rodríguez I, Franco-Navarro F, del Prado-Vera IC, Zavaleta-Mejia E. Control of *Nacobbus aberrans* in chili pepper (*Capsicum annuum* L.) by the combination of organic amendments, nematophagous fungi and nematicides. Nematropica. 2011;**41**(1):122-129

[95] Bascom-Slack CA, Arnold AE, Strobel SA. Student-directed discovery of the plant microbiome and its products. Science. 2012;**338**(6106):485-486

[96] Hallmann J, Quadt-Hallmann A, Mahaffee WF, Kloepper JW. Bacterial endophytes in agricultural crops. Canadian Journal of Microbiology. 1997;**43**(10):895-914

[97] Arnold A, Maynard Z, Gilbert G, Coley P, Kursar T. Are tropical endophytic fungi hyper diverse. Ecology Letters. 2000;**3**:267-274

[98] Aly AH, Debbab A, Kjer J, Proksch P. Fungal endophytes from higher plants: A prolific source of phytochemicals and other bioactive natural products. Fungal Diversity. 2010;**41**(1):1-16

[99] Quesada-Moraga E, Muñoz-Ledesma FJ, Santiago-Alvarez C. Systemic protection of *Papaver somniferum* L. against *Iraella luteipes* (Hymenoptera: *Cynipidae*) by an endophytic strain of *Beauveria bassiana* (Ascomycota: Hypocreales). Environmental Entomology. 2009;**38**(3):723-730

[100] Ownley B, Gwinn K, Vega F. Endophytic fungal entomopathogens with activity against plant pathogens: Ecology and evolution. BioControl. 2010;**55**:113-128

[101] Kauppinen M, Saikkonen K, Helander M, Pirttilä AM, Wäli PR. *Epichloë* grass endophytes in sustainable agriculture. Nature Plants. 2016;**2**:15224

[102] Vega F, Posada F, Aime M, Pava-Ripoll M, Infante F, Rehner S. Entomopathogenic fungal endophytes. Biological Control. 2008;**46**:72-82

[103] Jaber LR, Alananbeh KM. Fungal entomopathogens as endophytes reduce several species of *Fusarium* causing

crown and root rot in sweet pepper (*Capsicum annuum* L.). Biological Control. 2018;**126**:117-126

[104] Herre EA, Mejía LC, Kyllo DA, Rojas E, Maynard Z, Butler A, et al. Ecological implications of anti-pathogen effects of tropical fungal endophytes and mycorrhizae. Ecology. 2007;**88**(3):550-558

[105] Card S, Johnson L, Teasdale S, Caradus J. Deciphering endophyte behaviour: The link between endophyte biology and efficacious biological control agents. FEMS Microbiology Ecology. 2016;**92**(8):fiw114

[106] Hyde KD, Soytong K. The fungal endophyte dilemma. Fungal Diversity. 2008;**33**:163-173

[107] Shoresh M, Harman GE, Mastouri F. Induced systemic resistance and plant responses to fungal biocontrol agents. Annual Review of Phytopathology. 2010;**48**:21-43

[108] Klieber J, Reineke A. The entomopathogen *Beauveria bassiana* has epiphytic and endophytic activity against the tomato leaf miner *Tuta absoluta*. Journal of Applied Entomology. 2016;**140**(8):580-589

[109] Menjivar RD, Cabrera JA, Kranz J, Sikora RA. Induction of metabolite organic compounds by mutualistic endophytic fungi to reduce the greenhouse whitefly *Trialeurodes vaporariorum* (Westwood) infection on tomato. Plant and Soil. 2012;**352**(1-2):233-241

[110] Jaber LR, Ownley BH. Can we use entomopathogenic fungi as endophytes for dual biological control of insect pests and plant pathogens? Biological Control. 2018;**116**:36-45

[111] Lopez DC, Sword GA. The endophytic fungal entomopathogens *Beauveria bassiana* and *Purpureocillium lilacinum* enhance the growth of cultivated cotton (*Gossypium hirsutum*) and negatively affect survival of the cotton bollworm (*Helicoverpa zea*). Biological Control. 2015;**89**:53-60

[112] Gathage JW, Lagat ZO, Fiaboe KKM, Akutse KS, Ekesi S, Maniania NK. Prospects of fungal endophytes in the control of *Liriomyza leafminer* flies in common bean *Phaseolus vulgaris* under field conditions. BioControl. 2016;**61**(6):741-753

[113] Barra-Bucarei L, France Iglesias A, Gerding González M, Silva Aguayo G, Carrasco-Fernández J, Castro JF, et al. Antifungal activity of *Beauveria bassiana* endophyte against *Botrytis cinerea* in two *Solanaceae* crops. Microorganisms. 2020;**8**(1):65

3

Anthracnose of Chilli: Status, Diagnosis and Management

Raj Kiran, Jameel Akhtar, Pardeep Kumar and Meena Shekhar

Abstract

Chilli (*Capsicum annuum* L.) is one of the most economically important vegetable crops in the world. Among different biotic constraints, anthracnose disease is the major limiting factor affecting yield and production of chilli crop. Different symptoms associated with disease are fruit rot, leaf spots, dieback on stem, seedling blight, or damping off. Many species of genus *Colletotrichum* are found associated with the disease worldwide. In India, primarily three important species, namely, *Colletotrichum truncatum*, *C. acutatum*, and *C. gleosporoides*, are responsible for the chilli anthracnose. Accurate identification of pathogen is needed for choosing the proper management strategy for controlling this disease. Both conventional and molecular methods are adapted along with different management strategies, recommended for this disease namely cultural, chemical, and other eco-friendly methods.

Keywords: *Colletotrichum*, biocontrol agents, diagnosis, molecular methods

1. Introduction

The genus *Capsicum* includes many cultivated species, of which *Capsicum annuum* L. is one of the most widely cultivated one; besides this, other domesticated species are *C. baccatum*, *C. chinensis*, *C. frutescens*, and *C. pubescens* [1]. *C. annuum* comprises of both sweet (bell pepper) and pungent (chilli) fruits of numerous shapes and sizes. It is a good source of Vitamin A and C, potassium, and folic acid [2]. Fresh green chilli has more vitamin C than a citrus fruit, whereas red chilli has more vitamin A than in carrots [3, 4]. Besides its wide use as vegetable, spice, and condiments, it is also used in medicines and beverages. Capsaicinoid and caretenoids are the active ingredients of the chilli; the capsaicinoids are nonvolatile alkaloids that make chilli pungent [5], and caretenoids have nutritional value that also provides color to the chilli fruit [6]. In tropical and subtropical countries, chilli is considered the most important constituent of different cuisines. As the native home of chillies are considered to be tropical America, where it is still found growing in the wild state [7]. Its introduction to India is credited to voyage of Columbus who brought seeds from Spain, introducing it to Europe, which subsequently spread to Africa and Asia [8].

India is the world's largest producer of dried chillies and in 2018 India produced 1.8 million tons, out of 4.1 million tons produced worldwide [9]. There are two important commercial qualities that makes Indian chilli world famous are color and pungency levels. Chilli crop is attacked with different pests and pathogens in field and during post-harvest, contamination with mycotoxins are major constraints in chilli production. Worldwide, *Capsicum* is vulnerable to various pests, weeds, fungal, bacterial, and viral pathogens; among the fungal diseases, anthracnose/die-back/fruit-rot of chillies is an important disease causing serious losses in field, transit, transport, and storage [10, 11].

2. Anthracnose disease losses

The word anthracnose derived from Greek language meaning 'coal' it is the common name of plant disease with very dark, sunken lesions and containing fungal spores [12]. Typical symptoms (**Figure 1**) of anthracnose on chilli fruit include dark spots, sunken necrotic tissue with concentric rings of acervuli. Besides fruit rot, it also causes leaf spots, dieback on stem, seedling blight, or damping off. This disease not only affects the quality of fruit by appearance of anthracnose lesion but also reduces dry weight of fruit, and quantity of capsaicin and oleoresin [13, 14].

Losses are caused by this disease worldwide; it is reported that in Vietnam it causes 20–80% yield loss [15], 10% yield loss in Korea [16], 50% yield loss in Malaysia [17] and as high as 80% yield loss (during severe epidemics) in Thailand [18]. In India, a calculated loss of 10–54% has been reported in yield due to this disease [19, 20], and this disease is reported throughout India but it found to be more common and aggressive form in Assam, Bihar, Andhra Pradesh and Uttar Pradesh [10]. The anthracnose pathogen has been intercepted in seed and it has been reported that there is occurrence of pathogen in seed samples, upto 5% infection index indicates its wide spread occurrence in India [21].

Figure 1.
(a) Healthy chilli plant, (b) chilli plant affected with anthracnose disease, and (c) chilli fruits showing anthracnose symptom.

3. Causal organisms

This disease is caused by the species of genus *Colletotrichum*, which belongs to Ascomycetes. Worldwide, different species of *Colletotrichum* are reported to cause chilli anthracnose disease (**Table 1**), In India, among different species known to cause this disease, there are primarily three important species *Colletotrichum capsici* Syd. Butler and Bisby (Synonym *C. truncatum*), *C. acutatum* and *C. gloeosporioides* have been reported to be associated with the disease, however *C. truncatum* causing major damage at the ripe fruit stage of the plant [35, 52–55].

Country	Pathogen	References
Australia	*C. acutatum*, *C. atramentarium*, *C. dematium*, *C. gloeosporioides* var. minor and *C. gloeosporioides* var. *gloeosporioides* and *C. brisbanense*	[22, 23]
Brazil	*C. boninense*	[24]
India	*C. capsici/ C. dematium/ C. truncatum, C. gloeosporioides, C. graminicola, C. acutatum, C. piperatum, C. atramentaum, C. fructicola and C. siamense, C. cliviae, C. coccodes and C. karstii*	[20, 25–35]
Indonesia	*C. gloeosporioides*, *C. truncatum* and *C. acutatum*	[36]
South Korea	*C. acutatum*, *C. gloeosporioides*, *C. coccodes* and *C. dematium*	[37]
Mexico	*C. truncatum*	[23]
Malaysia	*C. truncatum*	[17, 38]
New Zealand	*C. coccodes*, *C. kartsii*, *C. novae-zelandiae* and *C. nigrum*	[39, 40]
Papua New Guinea	*C. truncatum* and *C. gloeosporioides*	[41]
Philippines	*C. gloeosporioides*, *C. truncatum* and *C. scovillei*	[42, 43]
Sri Lankla	*C. truncatum*	[44, 45]
Taiwan	*C. acutatum*, *C. truncatum* and *C. gloeosporioides*	[46]
Thailand	*C. acutatum*, *C. truncatum* and *C. gloeosporioides*	[47, 48]
USA	*C. gloeosporioides*, *C. acutatum*, *C. truncatum* and *C. coccodes*	[49, 50]
UK	*C. acutatum* and *Glomerella cingulata*	[51]
Vietnam	*C. acutatum*, *C. truncatum*, *C. gloeosporioides* and *C. nigrum*	[15]
Zimbabwe	*C. nymphaeae*	[23]

Table 1.
Different Colletotrichum species associated with the disease anthracnose of chilli in different countries.

4. Diagnosis

Identification of *Colletotrichum* species based on morphological characteris-tics (size and shape of conidia; presence of setae) and colony characteristics is generally used by several workers [56–59]; it is widely used in seed health testing labs for detection of *C. capsici* in germplasm for pest free conservation of chilli seeds [21]. As the pathogen is seed-borne, there is threat of introduction of this pathogen along with import of germplasm (including Chilli) from different countries; therefore, while importing from any other country, there is a need to examine the samples very critically including sensitive molecular diagnostic tools to prevent entry of this pathogen associated with germplasm [60]. Moreover, for the accurate identification of the pathogen at species level molecular methods are widely adapted. Loop-mediated isothermal amplification (LAMP) assay was used for the accurate and sensitive detection of *C. capsici* LAMP primers (β-tubulin gene sequences based) were designed and it was reported that it could detect as little as 10 fg/μl of *C. capsici* pathogen in comparison with only upto 1 ng/μl of *C. capsici* detection using polymerase chain reaction [61]. A sequence characterized amplified region (SCAR) marker was developed for specific and sensitive detection of *C. capsici* in chilli seeds and fruits. This markers did the amplification of an expected 250-bp fragment from genomic DNA and these markers were very much sensitive as it was reported that the marker could detect purified *C. capsici* DNA template up to 1 pg and DNA from *C. apsici* infected chilli fruits up to 25 ng [59]. As these two markers are very sensitive, these may be very useful in detection of the pathogen in imported germplasm in plant quarantine laboratories.

COL1/COL2 primers were used for amplification of the specific internal transcribed spacer region of tested *Colletotrichum* species (*C. acutatum*, *C. truncatum* and *C. gloeosporioides*) with a specific band of 460 base pairs. *C. gloeosporioides* was detected at a low level of 1000 conidia on chilli leaf and fruit by this primer [62]. Another, primer set based on the sequences of the ribosomal internal transcribed spacer (ITS1and ITS2) regions of *C. truncatum* was designed and standardized for the detection of *C. truncatum* in infected plant tissues using PCR assay. The sensitivity was 10 pg of genomic DNA from the pathogen [63]. Machenahalli et al. [64] detected pathogens from different parts of plant like seeds, fruits, infected twig/stem by PCR-based method by using specific primers. *C. truncatum* was amplified by spe-cies specific primer (C.cap-f and C.cap-r) as single band at 450 bp. *C. gloeosporioides* was amplified by species specific primers CgInt at 450 and *C. acutatum* by CaInt at 490 bp respectively. The accurate identification is very important for choosing the correct management strategy for this disease.

5. Management strategies

For the management of anthracnose disease of chilli, different strategies are adapted. These are use of cultural practices, chemical control, eco-friendly measures like use of biocontrol agents, plant extracts and use of resistant cultivars. Generally, use of different strategies in combination has been recommended for managing the disease [65]. The summarized information is given from across the world for the management of this disease.

5.1 Cultural practices

Several cultural practices have been reported to manage chilli anthracnose due to the special etiology of the pathogen. These precautionary measures are implemented to reduce the rate of infection and minimize the inoculum pressure even before fruits are mature and harvested. Than et al. [47] and Ali et al. [66] in their review reported that different cultural practices like disease free seeds, weeding, crop rotation, proper drainage, removal of crop residue are being followed for the chilli cultivation. It was suggested that disease free chilli seeds should be planted and elimination of weeds should be done in chilli field and rotation of chilli crop with other crops which are not alternative hosts to *Colletotrichum* spp. after every 2–3 years is very effective for controlling this disease. Good drainage systems on the field to channel out waste water during irrigation regimes, on-farm fruit disinfection such as fruit washing at packing houses and finally removal of plant debris which may serve as source of inoculum are some other clean crop and sanitation practices [47]. If there was history of disease in a particular field, then other crops should be rotated in isolation from other solanaceous plant for at least alternate years [50]. Deep plow is recommended to completely cover diseases plants or removing infected plant debris from the field at the end of growing season [67]. Early planting of chilli or planting cultivars that bear fruit within a short ripening period to allow the fruit to escape fungal infection is also recommended. Other alternative sanitation practices such as weeding, removal of infected or wounded fruits should be carried out regularly to prevent the pathogens from using such wounds as sites of infection.

5.2 Chemical control

Different strategies for managing the disease are recommended and chemical control is found most effective and practical method [68]. As time required for controlling the disease with chemical method is much lesser as compared to the time required for the development of resistant cultivar. Use of protective fungicide like manganese ethylene bisdithiocarbamate (Maneb) is widely recommended for managing this disease. Other dithiocarbamate fungicides like Mancozeb (0.2%), ziram (0.1%), copper oxychloride fungicide (Blitox 50), and Bordeaux mixture (0.5 or 1%) of a copper sulphate fungicide were found effective in managing this disease. Seed dressing with benzimidazole fungicides (Benlate, delsene M) and strobilurin fungicide (azoxystrobin) are recommended [69] and soaking of chilli seeds for 12 h in 0.2% Thiram, a dithiocarbamate fungicide was also found effective for better control of the disease [70].

Among different systemic fungicides recommended Bavistin (carbendazin 50% WP) 0.1%, Plantvax (oxycarboxin) and vitavax (carboxin) were found effec-tive as use of Bavistin resulted in 80.84% disease reduction [71] and Plantvax and Vitavax were reported to reduce the disease by checking the spore germination of *C. truncatum* [72]. Additionally other systemic fungicides from triazole group propiconazole [73], difenoconazole, benzimidazole fungicide (Benomyl) [74] have been used in both pre and post-harvest management of chilli anthracnose, as propiconazole, exhibited the highest level of inhibition of in vitro mycelial growth, biomass production, sporulation and spore germination at concentrations as low as 0.1 mg/ml. Other workers also reported that Tilt (propiconazole) is highly effective in controlling *Colletrotrichum* spp. [75, 76] concentration of Tilt at 150 ppm was found effective in inhibiting the pathogen as it caused 50% inhibition (ED 50) of *C. acutatum* growth in culture media [77]. It is to be noted that Benomyl and its associated fungicides Carbendazim and thiophanate methyl (both of which registered) has raised major health concerns and these are proved unacceptable and dangerous [78]. Different strobilurin fungicides azoxystrobin (Quadris), trifloxystrobin (Flint) and pyraclostrobin (Cabrio) have also been recommended for effective management of the disease [47, 79].

Moreover, dependence on only single chemical resulted in the emergence of resistant strains of *C. truncatum* isolates from chilli fruit against different chemicals benomyl, which were cross-resistant to thiophanate methyl and carbendazim [80], resistance of *C. truncatum* to benomyl and strobilurin-fungicides (azoxystrobin and kresoxim-methyl) is also reported [81–83]. Under such circumstances, combined application of Bioagents with chemicals are recommended, *Pseudomonas fluorescens* along with half of the recommended dose of azoxystrobin fungicide has been found effective and viable option to control fruit rot [79]. As use of chemicals are not eco-friendly and it leaves chemical residue in chilli fruits, which hinders the export, and there are numerous reports describing negative effects of using chemicals on farmer's health in developing countries [36]. To overcome the undesirable effects of chemical usage alternate methods such as use of bioagents, plant extracts or use of chemicals in combination with these are recommended to control the infection.

5.3 Biological control

Trichoderma species is the fungal antagonist which is widely applied to control *Colletotrichum* species in chilli [84, 85]. It is also believed that *Trichoderma* species are able to effectively compete for surface area, thereby reducing pathogen infection success [86–88]. Chloroform extracts of nonvolatile antibiotics (NVAC) of *T. viride* added to the culture media inoculated with *C. truncatum,* showed reduction in biomass and synthesis of RNA, DNA and protein [89]. It has been reported that antifungal metabolites (100 mg/L) secreted from *Trichoderma harzianum* Rifai strain number T-156co5 significantly controlled *C. truncatum* isolated from *C. annuum* [90]. *In vitro* studies indicated that *T. viride* and *P. fluorescens* are very effective in inhibiting mycelial growth of the pathogen [91]. It is suggested that the use of *T. viride* and *P. fluorescens* individually or in combination known to significantly lower the anthracnose disease incidence and should be used as an alternative to chemical control [92].

Other bioagents like *Bacillus subtilis* and *Candida oleophila* (a yeast species) have been tested for efficacy against *C. acutatum* [93]. *Pichia guilliermondii* Wick strain R13 is another yeast species which is reported to reduce the disease incidence on *C. truncatum* infected chilli fruit as low as 6.5%. It has also been proposed that this fungal strain with other yeasts suppressed *Colletotrichum* spp. through multiple modes of action (nutrient competition, competition for space between antagonist and the pathogen, toxin production, induction of plant resistance and hydrolytic enzyme production) [94–96]. Intanoo and Chamswarng [97] reported that DGg13 and BB133 were antagonistic bacterial strains found very effective in controlling *C. truncatum*. *Pseudomonas aeruginosa* FP6 also found effective against *C. acutatum* [98].

Rhizosphere and rhizoplane fungal isolates (*Chaetomium globosum*, *T. harzianum* and *F. oxysporum*) from perennial grasses has been reported to decreased disease incidence and severity in seedlings and mature plants, and promoted plant growth and increased yield in the greenhouse and field [99]. In an experiment crude extracts from *Chaetomium cupreum* CC, *C. globosum* CG, *T. harzianum* PC01, *T. hamatum* PC02, *Penicillium chrysogenum* KMITL44 and antibiotic substances Rotiorinol, Chaetoglobosin-C and Trichotoxin A50 was used against *C. gloeosporioides* isolate WMF01 (the most virulent on all tested varieties of grape). The results revealed that application of all bioproducts significantly reduced the disease incidence on leaves, twigs and fruits of grape in all varieties as compared to the chemical control [100]. *Cordyceps sobolifera* an entomopathogenic fungi have also been reported for use as a biocontrol agent against *C. gloeosporioides* [101, 102].

5.4 Plant extracts

Antimicrobial plant secondary metabolites compounds are one of the best options to controlling plant diseases. In chilli, several workers have shown the efficacy of plant extracts against *Colletotrichum* spp. [103–108]. Among the plant extracts, *Allium sativum* (10%) and *Azadirachta indica* (10%) demonstrated the highest inhibition of mycelial growth of *C. gloeosporioides* [91]. *A. indica, Datura stramonium, Ocimum sanctum, Polyalthia longifolia* and *Vinca rosea* were used against *C. truncatum*. Among the five fermented leaf extracts tested against *C. truncatum, A. indica extract* at 20% concentration highly inhibited the growth of *C. truncatum in vitro* condition. And *in vivo* the application of fermented leaf extract of *A. indica* alone reduced the fruit rot incidence (@3%) and increased plant height, number of fruits and yield significantly [109]. In an experiment the botanicals or plant extracts from *Catharanthus roseus*, *Coleus aromaticus*, *Manilka razapota* and *A. indica* used against fungi, it was concluded that these botanicals confer antifungal effects on the radial mycelial growth of *C. truncatum* [107]. The organic pesticides were prepared from the extract of neem leaves, soursop leaves, lemongrass extract, tuba root extract, and kenikir/*Cosmos caudate* extract [110]. The result indicates that neem leaves are the most effective organic pesticides to control the chilli pepper disease especially in Indonesia.

Nine plants extracts viz., *Lawsonia inermis, A. indica, Bougainvillea spectabilis, Withania sominifera, Ocimum tenuiflorum, Aegle marmelos* L., *Justicia adhatoda* and *Calotropis gigantean were* tested under *in vitro* condition through poisoned food technique against chilli fruit rot pathogen *Colletotrichum* sp., among them *W. sominifera* (10%) was found to highly inhibit the mycelial growth of the anthracnose pathogen up to 84.88% [111]. Further, Singh and Khirbat [112] reported the efficacy of aqueous extract of three wild plants viz., *Albizza lebbeck, Acacia arabica* and *Clerodendrum infortunatum* to control chilli fruit rot. Alves et al. [113] reported the efficacy of 1% aqueous or 20% ethanol plant extracts to control bell pepper anthrac-nose caused by *C. acutatum*. In this study, 6% aqueous garlic, mallow and ginger extracts reduced disease severity by more than 97%. Even though recent research suggests the use of these plant extracts as bio-fungicides, but still more studies on their efficacy in the controlling of chilli anthracnose need to be performed under field conditions.

5.5 Resistant cultivars

As use of resistant or tolerant cultivar is the most cost-effective management strategy. Due to the lack of resistance in the *C. annuum* gene pool, no commercial resistant varieties have been developed in *C. annuum* [114]. The introgression of the resistance gene from *C. baccatum* to *C. annuum* is difficult. There are some studies on introgression of anthracnose resistance into *C. annuum* to develop a new variety [115, 116]. Five lines of *C. annuum* from AVRDC, Taiwan, namely AVPP1102-B, AVPP0513, AVPP0719, AVPP0207 and AVPP1004-B, as the promising lines with good fruit yield and tolerance to anthracnose [117]. Two chilli varieties, Lembang-1 and Tanjung-2, have been reported as moderately resistant from IVEGRI, Indonesia, [118].

In India, some anthracnose-resistant lines listed are LLS, PBC932 (VI047018), Breck-2, PBC80 (VI046804), Breck-1, Jaun, and PBC81 (VI046805) [119]. Other nine resistant varieties (BS-35, BS-20, BS-28, Punjab Lal, Bhut Jolokia, Taiwan-2, IC-383072, Pant C-1 and Lankamura Collection) were identified which could be employed for developing successful resistant cultivars through breeding programs [120]. The information on the resistance varieties against *Colletotrichum* spp. may also be utilized for studying the inheritance of the resistance and also to locate and study the quantitative trait loci (QTLs) maps for resistance [121]. Further studies need to be undertaken to investigate the importance of these distinct genes in the management of chilli anthracnose. Nevertheless, the genetic mechanism associated with chilli resistance to anthracnose is still poorly understood mainly due to lack of information on the defense signaling modules governing the resistance mechanism.

6. Conclusion

Anthracnose of chilli is main constraint for its production in the India as well as worldwide. Detection this pathogen in the seed by the morphological features and with the developed molecular markers are very important especially in quarantine laboratories. The accurate detection of pathogen also helps in choosing the best management strategy for the control of this disease. Involvement of many *Colletotrichum* species in the disease and absence of resistance gene in *C. annum* makes breeding for resistance is more challenging. Moreover, injudicious use of chemicals for the control of this disease leaves residue in the chilli fruit poses threat to the export. Combining the use of resistance cultivars with other disease control measures would enhance the efficiency in integrated management of chilli anthracnose. Moreover, more research is required to find better alternative methods to control chilli anthracnose by involving vigorous evaluation and identification of resistant cultivars of chilli against this disease.

Author details

Raj Kiran, Jameel Akhtar*, Pardeep Kumar and Meena Shekhar
Division of Plant Quarantine, ICAR-National Bureau of Plant Genetic Resources, New Delhi, India

*Address all correspondence to: jameelnbpgr@gmail.com

References

[1] Tong N, Bosland PW. *Capsicum tovarii*, a new member of the *Capsicum baccatum* complex. Euphytica. 1999;**109**(2):71-77

[2] Pathirana R. Peppers: Vegetable and spice capsicums. In: Bosland PW, Votava EJ, editors. Crop Production Science in Horticulture Series 22. 2nd ed. Wallingford, UK: CAB International; 2012. p. 248

[3] Osuna-García JA, Wall MM, Waddell CA. Endogenous levels of tocopherols and ascorbic acid during fruit ripening of new Mexican-type Chile (*Capsicum annuum* L.) cultivars. Journal of Agricultural and Food Chemistry. 1998;**46**(12):5093-5096

[4] Marín A, Ferreres F, Tomás-Barberán FA, Gil MI. Characterization and quantitation of antioxidant constituents of sweet pepper (*Capsicum annuum* L.). Journal of Agricultural and Food Chemistry. 2004;**52**(12):3861-3869

[5] Díaz J, Pomar F, Bernal A, Merino F. Peroxidases and the metabolism of capsaicin in *Capsicum annuum* L. Phytochemistry Reviews. 2004;**3**(1-2):141-157

[6] Raghavan S. Handbook of Spices, Seasonings, and Flavorings. Boca Raton, USA: CRC; 2007. pp. 87-91

[7] MacNeish RS. Ancient mesoamerican civilization. Science. 1964;**143**(3606):531-537

[8] Heiser CB Jr. Peppers *Capsicum* (Solanaceae). In: Smartt J, Simmonds NW, editors. Evolution of Crop Plants. 2nd ed. UK: Wiley; 1995. pp. 265-268

[9] FAOSTAT. 2020. Available from: http://www.fao.org/faostat/en/#home

[10] Singh RS. Disease of Vegetable Crops. 2nd ed. New Delhi: Publishing Company Pvt. Ltd; 1987. p. 362

[11] Dev U, Akhtar J, Chaudhury R, Kandan A, Chand D, Kumar J, et al. Survival of *Colletotrichum capsici* (Syd.) Butler & Bisby in decade-long cryo-preserved chilli seeds. Seed Research. 2012;**40**(1):92-94

[12] Isaac S. Fungal-Plant Interactions. London: Chapman and Hall Press; 1992. p. 115

[13] Azad P. Fate and role of chemical constituents of chilli fruits during infection with *Colletotrichum capsici*. Indian Phytopathology. 1991;**44**:129-131

[14] Mistry DS, Sharma IP, Patel ST. Bio-chemical parameters of chilli fruits as influenced by *Colletotrichum capsici* (sydow) butler and bisby infection. Karnataka Journal of Agricultural Sciences. 2010;**21**(4):586-587

[15] Don LD, Van TT, Phuong Vy TT, Kieu PT. *Colletotrichum* spp. attacking on chilli pepper growing in Vietnam. Country report. In: Abstracts of the First International Symposium on Chilli Anthracnose. Republic of Korea: National Horticultural Research Institute, Rural Development of Administration; 2007. p. 24

[16] Byung SK. Country report of anthracnose research in Korea. In: First International Symposium on Chili Anthracnose. Seoul: Hoam Faculty House, Seoul National University; 2007. p. 24

[17] Sariah M. Incidence of *Colletotrichum* spp. on chili in Malaysia and pathogenicity of *C. gloeosporioides*. BIOTROP Special Publication. 1994;**54**:103-120

[18] Poonpolgul S, Kumphai S. Chilli pepper anthracnose in Thailand. In: Oh DG, Kim KT, editors. Abstracts of the First International Symposium on Chilli Anthracnose. Rural: National Horticultural Research Institute; 2007. p. 23

[19] Lakshmesha KK, Lakshmidevi N, Mallikarjuna SA. Changes in pectinase and cellulase activity of *Colletotrichum capsici* mutants and their effect on anthracnose disease on *capsicum* fruit. Archives of Phytopathology and Plant Protection. 2005;**38**(4):267-279

[20] Ramachandran N, Rathnamma K. *Colletotrichum acutatum* —A new addition to the species of chilli anthracnose pathogen in India. In: Annual Meeting and Symposium of Indian Phytopathological Society. Kasarago: Central Plantation Crops Research Institute; 2006. pp. 27-28

[21] Akhtar J, Singh B, Kandan A, Kumar P, Dubey SC. Status of *Colletotrichum* species infecting chilli germplasm processed for pathogen-free conservation in national gene bank, India. Bangladesh Journal of Botany. 2017;**46**(2):631-637

[22] Simmonds JH. Type specimens of *Colletotrichum gloeosporioides* var. minor and *Colletotrichum acutatum*. Queensland Journal of Agriculture and Animal Science. 1968;**25**:178

[23] Damm U, Woudenberg JH, Cannon PF, Crous PW. *Colletotrichum* species with curved conidia from herbaceous hosts. Fungal Diversity. 2009;**39**:45-87

[24] TozzeJr HJ, MassolaJr NM, Camara MP, Gioria R, Suzuki O, Brunelli KR, et al. First report of *Colletotrichum boninense* causing anthracnose on pepper in Brazil. Plant Disease. 2009;**93**(1):106. DOI: 10.1094/PDIS-04-12-0403-PDN

[25] Thind TS, Jhooty JS. Studies on variability in two *Colletotrichum* species causing anthracnose and fruit rot of chillies in Punjab. Indian Phytopathology. 1990;**43**(1):53-58

[26] Hegde GM, Anahosur KH, Srikant K. Biological control of *Colletotrichum capsici* causing fruit rot of chilli. Plant Pathology News. 2002;**20**:4-5

[27] Paul YS, Behl MK. Some studies on bell pepper anthracnose caused by *Colletotrichum capsici* and its control. Seed Science Research. 1990;**93**(1):656-659

[28] Susheela K. Evaluation of screening methods for anthracnose disease in chilli. Pest Management In Horticultural Ecosystems. 2012;**18**(2):188-193

[29] Selvakumar R. Variability among *Colletotrichum capsici* causing chilli anthracnose in north East India. In: Proceedings of the First International Symposium on Chilli Anthracnose, Hoam Faculty House, Seoul National University. Seoul, South Korea; 2007. p. 35

[30] Kaur S, Singh J. *Colletotrichum acutatum*-a new threat to chilli crop in Punjab. Indian Phytopathology. 1990;**43**(1):108-110

[31] Sharma PN, Kaur M, Sharma OP, Sharma P, Pathania A. Morphological, pathological and molecular variability in *Colletotrichum capsici*, the cause of fruit rot of chillies in the subtropical region of north western India. Journal of Phytopathology. 2005;**153**(4):232-237

[32] Sharma G, Shenoy BD. *Colletotrichum fructicola* and *C. siamense* are involved in chilli anthracnose in India. Archives of Phytopathology and Plant Protection. 2014;**47**(10):1179-1194

[33] Saini TJ, Gupta SG, Char BR, Zehr UB, Anandalakshmi R. First

report of chilli anthracnose caused by *Colletotrichum karstii* in India. New Disease Reports. 2016;**34**:6

[34] Krishnan S, Kaari M, Sawhney S, Sheoran N, Gautam RK, Das MM, et al. First report of *Colletotrichum siamense* from Andaman and Nicobar Islands causing anthracnose in chilli. Journal of Plant Pathology. 2019;**101**(3):767

[35] Akhtar J, Singh MK. Studies on the variability in *Colletotrichum capsici* causing chilli anthracnose. Indian Phytopathology. 2007;**60**(1):63-67

[36] Voorrips RE, Finkers R, Sanjaya L, Groenwold R. QTL mapping of anthracnose (*Colletotrichum* spp.) resistance in a cross between *Capsicum annuum* and *C. chinense*. Theoretical and Applied Genetics. 2004;**109**(6):1275-1282. DOI: 10.1007/s00122-004-1738-1

[37] Park KS, Kim CH. Identification, distribution and etiological characteristics of anthracnose fungi of red pepper in Korea. Journal of Plant Pathology. 1992;**8**:61-69

[38] Mahmodi F, Kadir J, Puteh A. Genetic diversity and pathogenic variability of *Colletotrichum truncatum* causing anthracnose of pepper in Malaysia. Journal of Phytopathology. 2013;**162**(7-8):456-465

[39] Damm U, Cannon PF, Woudenberg JH, Johnston PR, Weir BS, Tan YP, et al. The *Colletotrichum boninense* species complex. Studies in Mycology. 2012;**73**:1-36

[40] Liu F, Cai L, Crous PW, Damm U. Circumscription of the anthracnose pathogens *Colletotrichum lindemuthianum* and *C. nigrum*. Mycologia. 2013;**105**(4):844-860

[41] Pearson MN, Bull PB, Speke H. Anthracnose of *capsicum* in Papua New Guinea; varietal reaction and associated fungi. International Journal of Pest Management. 1984;**30**(3):230-233

[42] Opina NL. Occurrence, seed transmission and identification of *Colletotrichum* species causing pepper anthracnose in the Philippines and varietal screening for resistance. Philippine Phytopathology. 1993;**29**:72-83

[43] Cueva FD, Mendoza JS, Balendres MA. A new *Colletotrichum* species causing anthracnose of chilli in the Philippines and its pathogenicity to chilli cultivar Django. Crop Protection. 2018;**112**:264-268

[44] Rajapakse RG. Observations on anthracnose of chilli pepper *(Capsicum annuum L.)* caused by *Colletotrichum* species in Sri Lanka [doctoral dissertation]. Imperial College London (University of London); 1998

[45] Damm U, Cannon PF, Woudenberg JH, Crous PW. The *Colletotrichum acutatum* species complex. Studies in Mycology. 2012;**73**:37-113

[46] Manandhar JB, Hartman GL, Wang TC. Anthracnose development on pepper fruits inoculated with *Colletotrichum gloeosporioides*. Plant Disease. 1995;**79**:380-383

[47] Than PP, Jeewon R, Hyde KD, Pongsupasamit S, Mongkolporn O, Taylor PW. Characterization and pathogenicity of *Colletotrichum* species associated with anthracnose on chilli (*Capsicum* spp.) in Thailand. Plant Pathology. 2008;**57**(3):562-572. DOI: 10.1111/j.1365-3059.2007.01782.x

[48] Montri P, Taylor PW, Mongkolporn O. Pathotypes of *Colletotrichum capsici*, the causal agent of chili anthracnose, in Thailand. Plant Disease. 2009;**93**(1):17-20. DOI: 10.1094/PDIS-93-1-0017

[49] Harp TL, Pernezny K, Ivey ML, Miller SA, Kuhn PJ, Datnoff L. The etiology of recent pepper anthracnose

outbreaks in Florida. Crop Protection. 2008;**27**(10):1380-1384. DOI: 10.1016/ j. cropro.2008.05.006

[50] Roberts PD. Anthracnose caused by *Colletotrichum* sp. on pepper. University of Florida Cooperative Extension Service, Institute of Food and Agricultural Sciences, EDIS; 2001. Available from: http:// edis.ifas.ufl.edu/PP104

[51] Adikaram NK, Brown AE, Swinburne TR. Observations on infection of *Capsicum annuum* fruit by *Glomerella cingulata* and *Colletotrichum capsici*. Transactions of the British Mycological Society. 1983;**80**(3):395-401

[52] Ranathunge NP, Ford R, Taylor PW. Development and optimization of sequence tagged microsatellite site markers to detect genetic diversity within *Colletotrichum capsici*, a causal agent of chilli pepper anthracnose disease. Molecular Ecology Resources. 2009;**9**(4):1175-1179. DOI: 10.1111/ j.1755-0998.2009.02608.x

[53] Saxena A, Raghuwanshi R, Singh HB. Molecular, phenotypic and pathogenic variability in *Colletotrichum* isolates of subtropical region in north eastern India, causing fruit rot of chillies. Journal of Applied Microbiology. 2014;**117**(5):1422-1434. DOI: 10.1111/jam.12607

[54] Akhtar J, Khalid A. Virulence of *Colletotrichum capsici* and reaction on host differential. Journal of Mycology and Plant Pathology. 2008;**38**(3):683-684

[55] Akhtar J, Singh MK, Chaube HS. Effect of nutrition on formation of acervuli, setae and sporulation of the isolates of *Colletotrichum capsici*. Pantnagar Journal of Research. 2008;**6**(1):110-113

[56] Smith BJ, Black LL. Morphological, cultural, and pathogenic variation among *Colletotrichum* species isolated from strawberry. Plant Disease. 1990;**74**(1):69-76. DOI: 10.1094/ PD-74-0069

[57] Saxena A, Raghuwanshi R, Gupta VK, Singh HB. Chilli anthracnose: The epidemiology and management. Frontiers in Microbiology. 2016;**7**:1527

[58] Masoodi L, Anwar A, Ahmed S, Sofi TA. Cultural, morphological and pathogenic variability in *Colletotrichum capsici* causing die-back and fruit rot of chilli. Asian Journal of Plant Pathology. 2013;**7**(1):29-41

[59] Srinivasan M, Kothandaraman SV, Vaikuntavasan P, Rethinasamy V. Development of conventional and real-time PCR protocols for specific and sensitive detection of *Colletotrichum capsici* in chilli (*Capsicum annuum* L.). Phytoparasitica. 2014;**42**(4):437-444. DOI: 10.1007/s126 00-013-0380-3

[60] Singh B, Akhtar J, Kandan A, Kumar P, Chand D, Maurya AK, et al. Risk of pathogens associated with plant germplasm imported into India from various countries. Indian Phytopathology. 2018;**71**(1):91-102

[61] Kandan A, Akhtar J, Singh B, Pal D, Chand D, Agarwal PC, et al. Application of loop-mediated isothermal amplification (LAMP) assay for rapid and sensitive detection of fungal pathogen, *Colletotrichum capsici* in *Capsicum annuum*. Journal of Environmental Biology. 2016;**37**(6):1355-1360

[62] Imjit N, Rattanakreetakul C, Pongpisutta R. Polymerase chain reaction based detection of Chilli anthracnose disease. In: I International Conference on Postharvest Pest and Disease Management in Exporting Horticultural Crops. 2012. pp. 199-206

[63] Torres-Calzada C, Tapia-Tussell R, Quijano-Ramayo A, Martin-Mex R,

Rojas-Herrera R, Higuera-Ciapara I, et al. A species-specific polymerase chain reaction assay for rapid and sensitive detection of *Colletotrichum capsici*. Molecular Biotechnology. 2011;**49**(1):48-55

[64] Machenahalli S, Nargund VB, Hegde RV. Management of fruit rot causing seed borne fungal pathogens in chilli. The Bioscan. 2014;**9**(1):403-406

[65] Agrios GN. Plant Pathology. 3rd ed. New York: Academic Press; 1988. p. 845

[66] Ali A, Bordoh PK, Singh A, Siddiqui Y, Droby S. Post-harvest development of anthracnose in pepper (*Capsicum* spp): Etiology and management strategies. Crop Protection. 2016;**90**:132-141. DOI: 10.1016/j.cropro.2016.07.026

[67] Sharma M, Kulshrestha S. *Colletotrichum gloeosporioides*: An anthracnose causing pathogen of fruits and vegetables. Biosciences, Biotechnology Research Asia. 2015;**12**(2):1233-1246. DOI: 10.13005/bbra/1776

[68] Akhtar J. Effect of fungicides on growth and sensitivity of the isolates of *Colletotrichum capsici* causing fruit rot of chilli. Journal of Plant Disease Sciences. 2007;**2**(1):59-62

[69] Yu CH, Jin LH, Zhou MG. Effect of azoxystrobin on oxygen consumption and cyt b gene expression of *Colletotrichum capsici* from chilli fruits. Agricultural Sciences in China. 2009;**8**(5):628-631

[70] Chakravarthy BP, Anil Kumar TB. Control of seed borne infection of *Colletotrichum capsici* in chillies. Current Research. 1975;**4**:172

[71] Datar VV. Efficacy of growth regulators and fungitoxicants on fruit rot of chilli. Indian Journal of Mycology and Plant Pathology. 1996;**26**(3):239-242

[72] Narain A, Panigrahi C. Efficacy of some fungicidal compounds to control *Colletotrichum capsici* in vitro and in vivo. Indian Phytopathology. 1971;**24**:593-595

[73] Gopinath K, Radhakrishnan NV, Jayaraj J. Effect of propiconazole and difenoconazole on the control of anthracnose of chilli fruits caused by *Colletotrichum capsici*. Crop Protection. 2006;**25**(9):1024-1031. DOI: 10.1016/j.cropro.2006.02.001

[74] Boonyapipat P. Study of three fungicides to control anthracnose (*Colletotrichum capsici*) in chilli (*Capsicum frutescens*): Case study from Songkhla province, Thailand. In: I International Conference on Postharvest Pest and Disease Management in Exporting Horticultural Crops-(PPDM2012). 2012. pp. 103-108

[75] Smith BJ, Black LL. In vitro activity of fungicides against *Colletotrichum fragariae*. Acta Horticulturae. 1992;**348**:509-512

[76] Read PJ, Hide GA. Effects of fungicides on the growth and conidial germination of *Colletotrichum coccodes* and on the development of black dot disease of potatoes. The Annals of Applied Biology. 1995;**126**(3):437-447

[77] Freeman S, Nizani Y, Dotan S, Even S, Sando T. Control of *Colletotrichum acutatum* in strawberry under laboratory, greenhouse, and field conditions. Plant Disease. 1997;**81**(7):749-752

[78] APVMA. 2009. Australian pesticides and veterinary medicines authority, benomyl not registered or approved in Australia. Available from: https://archive.apvma.gov.au/archive/our_view/2009/2009-02-04_benomyl.php

[79] Anand T, Chandrasekaran A, Kuttalam S, Senthilraja G, Samiyappan R. Integrated control of fruit rot and powdery mildew of chilli using the biocontrol agent *Pseudomonas fluorescens* and a chemical fungicide. Biological Control. 2010;**52**(1):1-7. DOI: 10.1016/j.biocontrol. 2009.09.01 0

[80] Sariah M. Detection of benomyl resistance in the anthracnose pathogen, *Colletotrichum capsici*. Journal of Islamic Academy of Sciences. 1989;**2**(3):168-171

[81] Ramdial H, Rampersad SN. Characterization of *Colletotrichum* spp. causing anthracnose of bell pepper (*Capsicum annuum* L.) in Trinidad. Phytoparasitica. 2015;**43**(1):37-49. DOI: 10.1007/s12600-014-0428-z

[82] Inada M, Ishii H, Chung WH, Yamada T, Yamaguchi J, Furuta A. Occurrence of strobilurin-resistant strains of *Colletotrichum gloeosporioides* (*Glomerella cingulata*), the causal fungus of strawberry anthracnose. Japanese Journal of Phytopathology. 2008;**74**(2):114-117

[83] Hu MJ, Grabke A, Dowling ME, Holstein HJ, Schnabel G. Resistance in *Colletotrichum siamense* from peach and blueberry to thiophanate-methyl and azoxystrobin. Plant Disease. 2015;**99**(6):806-814. DOI: 10.1094/ PDIS-10-14-1077-RE

[84] Boonratkwang C, Chamswarng C, Intanoo W, Juntharasri V. Effect of secondary metabolites from *Trichoderma harzianum* strain Pm9 on growth inhibition of *Colletotrichum gloeosporioides* and chilli anthracnose control. In: Proceeding of the 8th National Plant Protection Conference. Phisanulok, Thailand: Naresuan University; 2007. pp. 323-336

[85] Singh MK, Akhtar J, Kumar A, Khalid A. Sensitivity of different isolates of *Colletotrichum capsici* to *Trichoderma* spp. Journal of Eco-friendly Agriculture. 2007;**2**(1):54-55

[86] Kashyap PL, Rai P, Srivastava AK, Kumar S. Trichoderma for climate resilient agriculture. World Journal of Microbiology and Biotechnology. 2017;**33**(8):155

[87] Maymon M, Minz D, Barbul O, Zveibil A, Elad Y, Freeman S. Identification of *Trichoderma* biocontrol isolates to clades according to ap-PCR and ITS sequence analyses. Phytoparasitica. 2004;**32**(4):370-3755

[88] Jeffries P, Koomen I. Strategies and prospects for biological control of diseases caused by *Colletotrichum*. In: Strategies and Prospects for Biological Control of Diseases Caused by *Colletotrichum*. Wallingford: Commonwealth Mycological Institute; 1992. pp. 337-357

[89] Rajathilagam R, Kannabiran B. Antagonistic effects of *Trichoderma viride* against anthracnose fungus *Colletotrichum capsici*. Indian Phytopathology. 2012;**54**:135-136

[90] Warin I, Chaiyawat S, Chiradej C, Montree I, Sorwaporn K, Kan C. Bioactive compound of antifungal metabolite from *Trichoderma harzianum* mutant strain for the control of anthracnose of chili (*Capsicum annuum* L.). Philippine Agricultural Scientist. 2009;**92**(4):392-397

[91] Ngullie M, Daiho L, Upadhyay DN. Biological management of fruit rot in the world's hottest chilli (*Capsicum chinense* Jacq.). Journal of Plant Protection Research. 2010;**50**(3):269-273

[92] Reddy YNP, Jakhar SS, Dahiya OS. Efficiency of bio-fungicides (*Trichoderma* spp and *Pseudomonas fluorescens*) on seedling emergence, vigour and health of infected Chilli seeds (*Capsicum annuum*) by

Colletotrichum capsici. Current Journal of Applied Science and Technology. 2019;**35**(5):1-8

[93] Wharton PS, Uribeondo JD. The biology of *Colletotrichum acutatum*. Anales Del Jardin Botanico De Madrid. 2004;**61**(1):3-22

[94] Janisiewicz WJ, Tworkoski TJ, Sharer C. Characterizing the mechanism of biological control of postharvest diseases on fruits with a simple method to study competition for nutrients. Phytopathology. 2000;**90**(11):1196-1200. DOI: 10.1094/PHYTO.2000.90.11.1196

[95] Chanchaichaovivat A, Panijpan B, Ruenwongsa P. Putative modes of action of *Pichia guilliermondii* strain R13 in controlling chilli anthracnose after harvest. Biological Control. 2008;**47**(2):207-215. DOI: 10.1016/j.biocontrol.2008.07.018

[96] Nantawanit N, Chanchaichaovivat A, Panijpan B, Ruenwongsa P. Induction of defense response against *Colletotrichum capsici* in chili fruit by the yeast *Pichia guilliermondii* strain R13. Biological Control. 2010;**52**(2):145-152

[97] Intanoo W, Chamswarng C. Effect of antagonistic bacterial formulations for control of anthracnose on chilli fruits. In: Proceeding of the 8th National Plant Protection Conference. Phisanulok, Thailand: Naresuan University; 2007. pp. 309-322

[98] Sasirekha B, Srividya S. Siderophore production by *Pseudomonas aeruginosa* FP6, a biocontrol strain for *Rhizoctonia solani* and *Colletotrichum gloeosporioides* causing diseases in chilli. Agriculture and Natural Resources. 2016;**50**(4):250-256. DOI: 10.1016/j.anres.2016. 02.003

[99] Vasanthakumari MM, Shivanna MB. Biological control of anthracnose of chilli with rhizosphere and rhizoplane fungal isolates from grasses. Archives of Phytopathology and Plant Protection. 2013;**46**(14):1641-1666. DOI: 10.1080/03235408.2013.771901

[100] Soytong K, Kanokmedhakul S, Rattanacherdchai K, Charoenporn C. Microbial elicitors to induce immunity for plant disease control in chilli and tomato. In: Basic and Applied Aspects of Biopesticides. New Delhi: Springer; 2014. pp. 99-125

[101] Imtiaj A, Lee TS. Screening of antibacterial and antifungal activities from Korean wild mushrooms. World Journal of Agricultural Sciences. 2007;**3**(3):316-321

[102] Jaihan P, Sangdee K, Sangdee A. Selection of entomopathogenic fungus for biological control of chili anthracnose disease caused by *Colletotrichum* spp. European Journal of Plant Pathology. 2016;**146**(3):551-564. DOI: 10.1007/s10658-016-0941-7

[103] Begum J, Yusuf M, Chowdhury JU, Khan S, Anwar MN. Antifungal activity of forty higher plants against phytopathogenic fungi. Bangladesh Journal of Microbiology. 2007;**24**(1): 76-78

[104] Nduagu C, Ekefan EJ, Nwankiti AO. Effect of some crude plant extracts on growth of *Colletotrichum capsici* (Synd) Butler & Bisby, causal agent of pepper anthracnose. Journal of Applied Biosciences. 2008;**6**(2):184-190

[105] Johnny L, Yusuf UK, Nulit R. Antifungal activity of selected plant leaves crude extracts against a pepper anthracnose fungus, *Colletotrichum capsici* (Sydow) butler and bisby (Ascomycota: Phyllachorales). African Journal of Biotechnology. 2011;**10**(20):4157-4165

[106] Saravanakumar P, Karthikeyan V, Patharajan S, Kannabiran B. Antifungal activity of Plumbago species against

anthracnose fungus *Colletotrichum gloeosporidodes* (Penz.) of chilli. Archives of Phytopathology and Plant Protection. 2011;**44**(3):287-297. DOI: 10.1080/03 235400903024803

[107] Ajith PS, Lakshmesha KK, Murthy SM, Lakshmidevi N. Botanicals for control of anthracnose of bell peppers. Journal of Plant Protection Sciences. 2012;**4**(1):13-19

[108] Rashid MM, Amin AR, Rahman F. Eco-friendly management of chilli anthracnose (*Colletotrichum capsici*). International Journal of Plant Pathology. 2015;**6**(1):1-11

[109] Nishanthi P, Gobika R, Charumathi M, Suji HA, Raj TS. Management of *Colletotrichum capsici* (SYD.) Butler and bisby causing fruit rot of chilli using fermented leaf extracts. Plant Archives. 2020;**20**(1):2509-2514

[110] Sopialena S, Sila S, Rosfiansyah R, Nurdiana J. The role of neem leaves as organic pesticides in chili pepper (*Capsicum frutescens*). Nusantara Bioscience. 2018;**10**(4):246-250

[111] Pavithra S, Akila R, Rajinimala N, GangaiSelvi R, Kannan R. Plant extraction mediated mitigation of chilli fruit rot caused by *Colletotrichum* spp. Journal of Pharmacognosy and Phytochemistry. 2019;**8**(4):2879-2883

[112] Singh K, Khirbat S. Inhibitory effects of leaf extracts of some plants on the fruit rot infecting *Capsicum annuum*. Annales Biologiques. 2014;**30**:365-367

[113] Alves KF, Laranjeira D, Câmara MP, Câmara CA, Michereff SJ. Efficacy of plant extracts for anthracnose control in bell pepper fruits under controlled conditions. Horticultura Brasileira. 2015;**33**(3):332-338

[114] Temiyakul P, Taylor PW, Mongkolporn O. Differential fruit maturity plays an important role in chili anthracnose infection. Journal King Mongkut's University of Technology North Bangkok. 2012;**22**(3):494-504

[115] AVRDC. Host resistance to pepper anthracnose. In: AVRDC Report. Shanhua, Taiwan: AVRDC—The World Vegetable Centre; 2003. pp. 29-30

[116] Pakdeevaraporn P, Wasee S, Taylor PW, Mongkolporn O. Inheritance of resistance to anthracnose caused by *Colletotrichum capsici* in capsicum. Plant Breeding. 2005;**124**(2):206-208

[117] Hasyim A, Setiawati W, Sutarya R. Screening for resistance to anthracnose caused by *Colletotrichum acutatum* in chili pepper (*Capsicum annuum* L.) in Kediri, East Java. Advances in Agriculture & Botanics. 2014;**6**(2):104-118

[118] Setiawati W, Udiarto BK, Soetiarso TA. The effect of variety and planting system of chili pepper on incidence of whiteflies. The Horticulture Journal. 2008;**18**(1):55-61

[119] Reddy MK, Srivastava A, Kumar S, Kumar R, Chawda N, Ebert AW, et al. Chilli (*Capsicum annuum* L.) breeding in India: An overview. SABRAO Journal of Breeding and Genetics. 2014;**46**(2):160-173

[120] Garg R, Loganathan M, Saha S, Roy BK. Chilli anthracnose: A review of causal organism, resistance source and mapping of gene. In: Kharwar RN, Upadhyay RS, Dubey NK, Raghuwansh R, editors. Microbial Diversity and Biotechnology in Food Security. India: Springer; 2014. pp. 589-610

[121] Mahasuk P, Struss D, Mongkolporn O. QTLs for resistance to anthracnose identified in two *Capsicum* sources. Molecular Breeding. 2016;**36**(1):10. DOI: 10.1007/s11032-016-0435-5

4

Capsicum: Chemistry and Medicinal Properties of Indigenous Indian Varieties

Mithun Rudrapal and Khomendra Kumar Sarwa

Abstract

Capsicum pods (or chili pepper) constitute the world's second most consumed vegetable crop and spice after tomato in our daily culinary practice. Five indigenous species that are widely domesticated in various parts of India include *Capsicum annuum* L., *Capsicum chinense* Jacq., *Capsicum frutescens* L., *Capsicum baccatum* L., and *Capsicum pubescens* L. The chili pods of *C. chinense* Jacq (locally known as *Bhut Jolokia* in the Assam state of India) was officially recognized as the world's hottest *Capsicum* variety according to the Guinness Book of World Records. Capsaicinoids, a group of chemical principles present in matured *capsicum* pods, are responsible for the pun-gency as well as pharmacological/medicinal properties of capsicum. Some important capsaicinoids include capsaicin, followed by dihydrocapsaicin, nordihydrocapsaicin, homodihydrocapsaicin, and homocapsaicin. Traditional practices of capsicum pods of *Bhut Jolokia* have been well documented for the management of various human disor-ders/ailments, particularly in the north-eastern region of India. In modern medicine, *Bhut Jolokia* pods have been used in the treatment of arthritis, gastritis, toothache, and musculoskeletal and neuropathic pain including other pharmacological disorders and microbial infections. Capsaicinoids have been reported to exhibit a diverse range of biological effectiveness such as antioxidant, analgesic, and anti-inflammatory, anticarcinogenic effects.

Keywords: *capsicum*, capsaicinoids, capsaicin, *Bhut Jolokia*, *Dhan Jolokia*, indigenous, Indian varieties, pungency, traditional uses, medicinal properties

1. Introduction

Capsicum (also called as Peeper or Chili pepper) is an important vegetable crop widely cultivated in tropical and subtropical regions of the world. This spice crop has been well known since past 9500 years. *Capsicum* pods or chili pepper constitute the world's second most consumed vegetable crop and spice after tomato in our daily culinary practice [1].

India is the largest producer, consumer, and exporter of *Capsicum* in the world. It contributes about 36% to global production of *Capsicum* and exports about 20% of its total production [2]. The production of *Capsicum* in India is dominated by the state of Andhra Pradesh, which contributes 53% of the total production followed by Karnataka (9%), Odisha (6%), West Bengal (6%), Maharashtra (5%), and Madhya Pradesh (4%). *Capsicum* is also cultivated in other parts of India including the north-eastern states [3].

Several indigenous varieties of *Capsicum* are known to be cultivated in the north-eastern region of India. Some of such indigenous varieties have been well documented to be the hottest *Capsicum* varieties in the world [4]. These indigenous varieties are popularly known by different local names among various communities of the north-eastern states of India. These are namely, *Bih Jolokia* (*Bih* means poison, *Jolokia* means pepper), *Bhut Jolokia* (Bhut means ghost) in Assamese language, *Oo-Morok* (*Oo* means Tree, *Morok* means Chili) in Manipuri language, and *Naga Jolokia* (or *Naga Morich*) in Nagamese language and *Raja Chili* (King of Chili). The word *Jolokia* usually refers to the vernacular name of *Capsicum* pod or chili pepper in Assamese language of India [5–7].

2. Indigenous *Capsicum* varieties of India

The plant of *Capsicum* belongs to the genus *Capsicum*, which is a member of the family Solanaceae [8]. The genus *Capsicum* has approximately 27 species, out of which 5 species are widely domesticated in various parts of India. These are namely, *Capsicum annuum* L., *Capsicum chinense* Jacq., *Capsicum frutescens* L., *Capsicum baccatum* L., and *Capsicum pubescens* L. [9]. The images of indigenous *Capsicum* varieties of India are presented in **Figure 1**.

Capsicum annuum is the most cultivated species across different parts of India. *Capsicum chinense* and *Capsicum frutescens* are also the most common cultivated species, particularly in the north-eastern region of India. Other three species including *Capsicum baccatum* and *Capsicum pubescens* are also found to be cultivated in the north-eastern states of India [10]. **Table 1** describes five indigenous *Capsicum* varieties of India.

Figure 1.
Photographs showing indigenous Capsicum *varieties of India.*

Species/botanical name	*Capsicum* variety/ vernacular name	Description
Capsicum annum	*Jati Jolokia*	Elongated, 2.5–3.0 cm long, width 0.8–1.0 cm, smooth surface, light red, characteristic odor
Capsicum chinense	*Bhut Jolokia*	Elongated, 5.0–7.0 cm long, 2.53.0 cm width, undulating rough surface, dark red, characteristic aroma
Capsicum frutescens	*Dhan Jolokia*	Tiny in size, elongated, 1.0–1.5 cm long, 0.5 cm wide, smooth surface, light red, characteristic aroma
	Maam Jolokia	Elongated, 1.5–1.75 cm long, width 0.5–0.75 cm, smooth surface, light red, characteristic aroma
	Totta Bias	Elongated, 4.0–5.0 cm long, 1.0 cm wide, smooth surface, orange to light red in color
Capsicum baccatum	*Ohm Jolokia*	Rough shape with a flat base, 2.0–3.0 cm diameter, 2.0–2.5 height, and base divided into three lobes, each having 3–4 grooves, dark red
Capsicum pubescens	*Bhikue Jolokia*	Bell shaped like a tomato with a flat base, 2.0–2.2 cm diameter, 1.5 cm height, smooth surface, dark red, characteristic aroma

Table 1.
Indigenous Capsicum *varieties of India [10, 11].*

3. Chemistry and pungency

The pungency of *Capsicum* pods is mainly attributed to the presence of a group of closely related compounds called capsaicinoids. Capsaicinoids refer to a group of pungent alkaloids that accumulate in the placenta of matured *Capsicum* pods [12]. Chemically, these are acid amides of vanillylamine with C9–C12 branched-chain fatty acid. All capsaicinoids possess a common 3-hydroxy-4-methoxybenzylamide (vanilloid) skeleton, but differ in their hydrophobic alkyl side chain. Differences in the side chain include saturation of the carbon carbon double bond, absence of a methyl group and also changes in the length of hydrocarbon chain [13]. Some important capsaicinoids are capsaicin, followed by dihydrocapsaicin, nordihydrocapsaicin, homodihydrocapsaicin, and homocapsaicin. Capsaicin and dihydrocapsaicin together account for approximately 80% of total capsaicinoids content of *Capsicum* pods. These two compounds are the most important and potent members of capsaicinoids family [14]. The chemical structures of capsaicinoids are displayed in **Figure 2**.

Most of the *Capsicum* species cultivated in India contain around 1% of capsaicinoids, but *Capsicum chinense* L. (*Bhut Jolokia*) and *Capsicum frutescens* (*Dhan Jolokia*) possess around 2–4% of capsaicinoids [15]. However, though seven different indigenous varieties of *Capsicum* are found to be cultivated in the Assam state of India, the *Bhut Jolokia* and *Dhan Jolokia* have been documented to be the most potential *Capsicum* varieties in terms of their capsaicinoids content and level of pungency [16]. The capsaicinoids content (%) of indigenous *Capsicum* varieties of India is depicted in **Table 2**.

It is noteworthy that the capsaicinoids content is a measure of the degree of hotness or the level of pungency of *Capsicum* pods. However, it is also responsible for medicinal properties as well as for the nutritional value of *Capsicum* spice. The Defense Research Laboratory, Tezpur, India reported that the potential of the *Bhut*

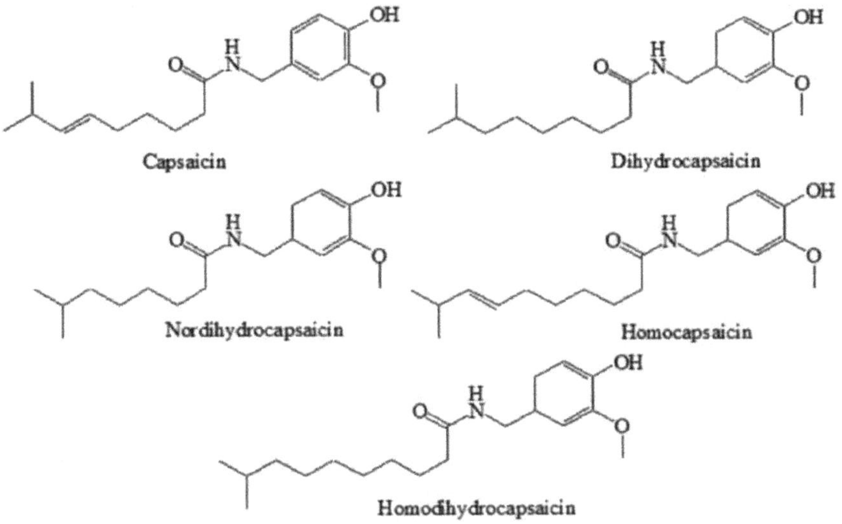

Figure 2.
Chemical structures of capsaicinoids.

Species	*Capsicum* variety	Capsaicinoids content (%)
Capsicum annum	*Jati Jolokia*	0.51
Capsicum chinense	*Bhut Jolokia*	2.45
Capsicum frutescens	*Dhan Jolokia*	2.14
	Maam Jolokia	1.38
	Totta Bias	0.25
Capsicum baccatum	*Ohm Jolokia*	0.67
Capsicum pubescens	*Bhikue Jolokia*	0.92

Table 2.
Capsaicinoids content (%) of indigenous Capsicum *varieties of India [15, 16].*

Jolokia is comparable with that of the Red Savina Habanero of Mexico in terms of Scoville Heat Units (SHUs) [17]. The pungency of *Bhut Jolokia* was recorded 855,000 SHU, while it was 577,000 SHU for Red Savina Habanero. Studies at the New Mexico State University also reported that *Bhut Jolokia* possessed 1,001,304 SHUs, whereas Red Savina had 248,556 SHUs [18]. In September 2006, the *Bhut Jolokia* was officially recognized as the world's hottest *Capsicum* variety measuring over 1,000,000 SHUs according to the Guinness Book of World Records [19].

4. Medicinal properties

Apart from wide commercial applications of *Capsicum* pods or chili pepper as culinary spice and also their use in various food products such as seasoning blends and in the canning industry, *Capsicum* also possess a diverse range of medicinal and/or pharmacological potential. In medicine, capsaicinoids, the active ingredients of *Capsicum* pods, have been used for the treatment of gastritis, arthritis, toothache, musculoskeletal and neuropathic pain, chronic indigestion, other pharmacological disorders and also microbial infections [20–22]. Capsaicinoids have been reported to exhibit a diverse range of biological effectiveness such as antioxidant, analgesic, anti-inflammatory, anticarcinogenic, promotion of energy metabolism, and suppression of fat accumulation [23–26].

Traditional practices of *Bhut Jolokia* have been well documented for the management of various human disorders/ailments, particularly in the north-eastern region of India. It has also been found to be effective in tuning up body muscles and in the treatment of wrist pain after a strenuous physical exercise [5]. The treatment of toothache and muscle pain using the hot infusion of *C. chinense* pods has also been reported in other parts of India [27]. Traditional uses of *C. chinense* leaves in several human aliments such as boil, headache, and night blindness have been well documented, for example, the use of leaf paste in the treatment of boils [28]. The use of *Bhut Jolokia* has also been found effective in the symptomatic relief of asthma because of its bronchodilation effect. Regular consumption of small quantities of *Capsicum* pods of *Bhut Jolokia* is highly beneficial in certain gastrointestinal abnormalities [5, 29].

Researchers have reported that the capsaicinoids content might be responsible for the pharmacological and/or biological potential of *Capsicum* pods. In modern medicine, the capsaicin has been considered to be an effective yet safe topical analgesic as antiarthritic, antioxidant, and anticancer agent. The antiviral efficacy of capsaicin has been reported in the treatment of herpes zoster infection [30]. The Osteoarthritis Research International (ORI) recommended the topical use of capsaicin as an effective adjunctive or alternative medicament to oral analgesic/anti-inflammatory agent for the treatment of moderate to severe pain and inflammation in case where conventional oral analgesic/anti-inflammatory agents generally do not respond [31].

As mentioned above, capsaicin has, therefore, been included in topical therapy for the relief of different neuropathic pain, although it could produce skin irritation. Capsaicin and its analogues have been used in certain topical dosage forms such as creams and patches in order to treat chronic pain syndromes such as post-herpetic neuralgia, musculoskeletal pain, diabetic neuropathy, osteoarthritis, and rheumatoid arthritis [32, 33]. It has also been found beneficial in relieving pain due to rashes, psoriasis, mastectomy, and bladder disorders. Adverse effects (burning, stinging and erythema) are normally limited to the site of application; however, respiratory irritations and occasional systemic effects arising from the inhalation of cream have been reported [34].

5. Mechanism of action

The mechanism of analgesia that capsaicinoids produce is probably due to their counter-intuitive effect [35]. Capsaicinoids applied directly to the skin or injected by intradermal route induce both hyperalgesia and allodynia. Capsaicin causes excitation of certain subsets of dorsal root ganglia giving rise to the stimulation of thin myelinated A-delta and unmyelinated C fibers. These nerve fibers are specific for the transmission of pain signals and therefore seem to be sensitive to vanilloids such as capsaicin, which blocks particularly the transmission of noxious stimuli [36]. Due to the blockade of terminal peripheral nerves, capsaicin inhibits the release of several pro-inflammatory neuropeptides such as Substance P, calcitonin gene-related peptide, and somatostatin [37]. The prolonged stimulation of neurons by capsaicin may deplete the nerve's ability to release the presynaptic neuropeptides, particularly Substance P. The ability of capsaicin to desensitize nociceptors is the main reason behind its therapeutic efficacy [38]. It has also been attributed that capsaicin could modulate the amount of specific neurotransmitter called Substance P, associated with the feeling of pain and thus helps alleviate pain externally [39]. Research findings claim that pure capsaicin could possess similar efficacy in treating joint and muscle pain induced in rheumatism, arthritis, ankylosing spondylitis, and fibromyalgia. Further, clinical reports suggest that capsaicin increases not only the secretion of saliva and gastric juice, but also improves the blood flow in the lining of the gastrointestinal tract, which might be beneficial in gastrointestinal abnormalities [40, 41].

6. Conclusion

Based upon ethnomedicinal potential and scientific evidences of possessing extreme pungency, *Capsicum* chili pod of *Bhut Jolokia* (*Capsicum chinense* Jacq.) is an important economically viable crop of north-eastern region of India. Further, it is believed that favorable geographical and agroclimatic factors are primarily responsible for the higher capsaicinoids content of *Capsicum*. The functional quality depends primarily on the capsaicinoids content, which determines the medicinal potential and/or biological efficacy of *Capsicum* pods. However, *Capsicum* chili pods of *Bhut Jolokia* may be studied for further scientific investigation toward their possible development into topically effective analgesic drugs and/or formulations for the treatment of pain-related disorders and inflammatory illness. In addition to *Bhut Jolokia,* another indigenous variety of *Capsicum*, that is, *Dhan Jolokia* (*Capsicum frutescens*) may also be studied for further scientific investigations.

Author details

Mithun Rudrapal[1*] and Khomendra Kumar Sarwa[2]

1 Sandip Institute of Pharmaceutical Sciences, Nashik, India

2 Department of Pharmacy, Govt. Girl's Polytechnic, Raipur, India

*Address all correspondence to: rsmrpal@gmail.com

References

[1] Sreedhara DS, Kerutagi MG, Basavaraja H, Kunnal LB, Dodamani T. Economics of *Capsicum* production under protected conditions in Northern Karnataka. Journal of Agricultural Science. 2013;**26**(2):217-219

[2] Yoon JY, Green SK, Tshanz AJ, Tsou SCS, Chenge LC. Pepper improvement for the tropics problems and approach. In: Tomato and Pepper Production in the Tropics. Taiwan: AVRDC Shantine, Asian Vegetable Research and Development Center; 1989. pp. 86-90

[3] Spices Board, Government of India. Available from: http://www.indianspices.com

[4] Bosland PW, Baral JB. *Bhut Jolokia*—The world's hottest known chili pepper is a putative naturally occurring interspecific hybrid. Horticultural Science. 2007;**42**:222-224

[5] Bhagowati RR, Changkija S. Genetic variability and traditional practices in naga king chili landraces of Nagaland. Asian Agri-History Foundation. 2009;**13**:171-180

[6] Sarwa KK, Mazumder B, Rudrapal M, Debnath M, Kumar A, Verma VK, et al. Capsaicinoids content of some indigenous *Capsicum* varieties of Assam, India. Journal of Natural Science Research. 2013;**3**(4):112-116

[7] Sarwa KK, Mazumder B, Rudrapal M. Some indigenous *Capsicum* species of India and their ethnomedicinal importance. In: Kumar S, editor. Recent Advances in Ethnobotany. New Delhi: Deep Publishers; 2015. pp. 18-21

[8] Sarwa KK, Kiran K, Sahu J, Rudrapal M, Debnath M. A short review on *Capsicum Chinense* Jacq. Journal of Herbal Medicine & Toxicology. 2012;**6**(2):7-10

[9] Vidhi J. Capsicum: Origin, Flower Structure and Varieties, India. Available from: http://www.biologydiscussion.com

[10] Vidhi J. Main Types of Capsicum, India. Available from: http://www.biologydiscussion.com

[11] Antonious GF, Jarret RL. Screening capsicum accessions for capsaicinoids content. Journal of Environmental Science and Health (Part B). 2006;**41**(5):717-729

[12] Wesołowska A, Jadczak D, Grzeszczuk M. Chemical composition of the pepper fruit extracts of hot cultivars *Capsicum annuum* L. Acta Scientiarum Polonorum Hortorum Cultus. 2011;**10**:171-184

[13] Luo XJ, Peng J, Li YJ. Recent advance in the study on capsaicinoids. European Journal of Pharmacology. 2011;**650**:1-7

[14] Estrada B, Bernal MA, Díaz J, Pomar F, Merino F. Capsaicinoids in vegetable organ of *Capsicum annum* L. in relating to fruiting. Journal of Agricultural and Food Chemistry. 2002;**50**(5):1188-1191

[15] Sarwa KK, Mazumder B, Rudrapal M. Effect of extraction methods and long term storage on capsaicinoids content of *Bhut Jolokia* fruits. Indian Journal of Natural Products and Resources. 2017;**8**(1):69-77

[16] Sanatombik K, Sharma GJ. Capsaicin content and pungency of different *Capsicum* spp. cultivars. Notulae Botanicae Horti Agrobotanici Cluj-Napoca. 2008;**36**(2):89-90

[17] Mathur R, Dangi RS, Dass SC, Malhotra RC. The hottest chilli variety in India. Current Science. 2000;**79**:287-288

[18] Scoville WL. Note on *Capsicum*. Journal of the American Pharmaceutical Association. 1912;**1**:453-454

[19] Guinness Book of World Records. Hottest Spice. 2006. Available from: http://www.guinnessworldrecords.com

[20] Jolayemi AT, Ojewole JAO. Comparative anti-inflammatory properties of Capsaicin and ethyl-acetate extract of *Capsicum frutescens* Linn [Solanaceae] in rats. African Health Sciences. 2013;**13**(2):357-361

[21] Materska M, Perucka I. Antioxidant activity of the main phenolic compounds isolated from hot pepper fruit (*Capsicum annuum* L.). Journal of Agricultural and Food Chemistry. 2005;**53**:1750-1756

[22] Molina-Torres J, Garcia-Chavez A, Ramirez-Chavez E. Antimicrobial properties of alkamides present in flavouring plants traditionally used in Mesoamerica: Affinin and capsaicin. Journal of Ethnopharmacology. 1999;**64**:241-248

[23] Clark R, Lee SH. Anticancer properties of capsaicin against human cancer. Anticancer Research. 2016;**36**(3):837-843

[24] Rosa A, Deiana M, Casu V, Paccagnini S, Appendino G, Ballero M, et al. Antioxidant activity of capsaicinopids. Journal of Agricultural and Food Chemistry. 2002;**50**(25):7396-7401

[25] Srinivasan K. Biological activities of red peeper (*Capsicum annum*) and its pungent principle capsaicin: A review. Critical Reviews in Food Science and Nutrition. 2016;**56**(9):1488-1500

[26] Janssens PLHR, Hursel R, Martens EAP, Westerterp-Plantega MS. Acute effects of capsaicin on energy expenditure and fat oxidation in negative energy balance. PLoS One. 2013;**8**(7):e67786

[27] Rout SD, Panda SK. Ethnomedicinal plant resources of Mayurbhanj district, Orissa. Indian Journal of Traditional Knowledge. 2010;**9**:68-72

[28] Liu Y, Nair MG. Capsaicinoids in the hottest pepper *Bhut Jolokia* and its antioxidant and anti-inflammatory activities. Natural Product Communications. 2010;**5**:91-94

[29] Whiting S, Derbyshire E, Tiwari BK. Capsaicinoids and capsinoids. A potential role for weight management? A systematic review of the evidence. Appetite. 2012;**59**(2):341-348

[30] Fusco MB, Giacovazzo M. Peppers and pain: The promise of capsaicin. Drugs. 1997;**53**:909-914

[31] Zhang W, Moskowitz RW, Nuki G, Abramson S, Altman RD, Arden N, et al. OARSI recommendations for the management of hip and knee osteoarthritis, Part II: OARSI evidence based, expert consensus guidelines. Osteoarthritis and Cartilage. 2008;**16**:137-162

[32] Backonja MM, Malan TP, Vanhove GF, Tobias JK. NGX-4010, a high-concentration capsaicin patch, for the treatment of postherpetic neuralgia: A randomized, double-blind, controlled study with an open-label extension. Pain Medicine. 2010;**11**:600-608

[33] Tesfaye S. Advances in the management of diabetic peripheralneuropathy. Current Opinion in Supportive and Palliative Care. 2009;**3**:136-143

[34] Sawynok J. Topical analgesics in neuropathic pain. Current Pharmaceutical Design. 2000;**11**:2995-2300

[35] Reyes-Escogido ML, Gonzalez-Mondragon EG, Vazquez-Tzompantzi E. Chemical and pharmacological aspects of capsaicin. Molecules. 2011;**16**:1253-1270

[36] Carte RB. Topical capsaicin in the treatment of cutaneous disorders. Drug Development Research. 1999;**22**:109-123

[37] Derry S, Lloyd R, Moore RA, McQuay HJ. Topical capsaicin for chronic neuropathic pain in adults. Cochrane Database of Systematic Reviews. 2009;**7**(4):CD007393

[38] Prausnitz MR, Mitragotri S, Langer R. Current status and future potential of transdermal drug delivery. Nature Reviews. Drug Discovery. 2004;**3**(2):115-124

[39] Casale R, Symeonidou Z, Bartolo M. Topical treatments for localized neuropathic pain. Current Pain and Headache Reports. 2017;**21**(3):15

[40] Sarwa KK, Mazumder B, Rudrapal M. Topical ethosomal capsaicinoids attenuates edema and nociception in arthritic rats. Drug Delivery. 2015;**22**(8):1043-1052

[41] Sarwa KK, Mazumder B, Rudrapal M, Verma VK. Potential of capsaicin loaded transferosomes in arthritic rats. Drug Delivery. 2015;**22**(5):638-646

5

Hot Water Seed Treatment: A Review

Suryapal Singh, Harshita Singh and Narender K. Bharat

Abstract

In present day agriculture, use of chemicals for crop production is discouraged. Hence, other alternative treatments for disease control must be developed, and hot water treatment is one of them. It is a feasible practice, both financially and time wise. Hot water soaking is a very age-old practice, efficient in destroying pathogens borne both outside the testa and inside the seed testa by using temperature hot enough to kill the organism but not quite hot enough to kill the seed. Extensive research work has been reported on hot water treatment in vegetables. Therefore, an attempt has been made to review the information available regarding the effect of hot water treatment on growth, disease incidence and yield of vegetables.

Keywords: hot water treatment, capsicum, seed-borne disease

1. Introduction

In the present age when the crop production is protected by chemicals and threat of disease due to exposure to these chemicals is at a rise, organic farming is getting momentum. This organic farming is regulated by certain certifications including chemical-free state of agriculture farm. In this condition use of chemicals will not help farmers for these certifications; rather, there is rejection of the produce. Here it becomes pertinent to involve those practices which are natural and can help control diseases naturally. Priming and pelleting are commonly used practices for seed treatment to enhance the production of *Capsicum*, but some chemicals are involved in it. *Capsicum* is in great demand as it is a vegetable with very rich contents of vitamins and minerals and is cooked in combination and individually as per the taste of consumers. Rejections of pesticide, insecticide and chemical sprayed crop attracted the attention for the present review where hot water seed treatment is controlling the diseases in bell pepper. Pepper (*Capsicum annuum* L. var. *grossum*) is commonly known as *Capsicum*, *Shimla Mirch*, green pepper, cherry pepper or bell pepper, and it belongs to the solanaceous group of vegetables. *Capsicum* is one of the most important vegetable crops grown extensively throughout the world especially in the temperate countries [1] and has attained a status of high value crop in recent years because of its delicacy and pleasant flavour coupled with rich content of ascorbic acid and other vitamins and minerals [2]. Fruits of sweet pepper are either used as salad or cooked as vegetable or processed and is appreciated worldwide for its flavour, aroma and colour. It contains several bioactive substances like capsaicin, vitamin E, pro-vitamin A, carotenoids, flavonoids and other secondary metabolites with antioxidant properties [3]. In the states of Arunachal Pradesh, Himachal Pradesh, Jammu and Kashmir, Darjeeling District of West Bengal and hills of UP, bell pepper is grown in summer, whereas in the states of Karnataka, Tamil Nadu, Bihar and Maharashtra, it is grown in the autumn season [4]. In India, bell pepper covers an area of 29,800 ha with 171,370 tonnes production and productivity of 5.75 tonnes/ha [5]. Sweet pepper is a warm-season crop, which grows well under an extended frost-free season, with the capability of producing high yields of exceptional quality. The best temperature range for sweet pepper growth is 20–25°C, whereas the best germination temperature is 29°C. A temperature of less than 29°C reduces the growth of seedling, leading to increased exposure of seedlings to insects, diseases or salts, resulting in damaged or dead seedlings [6]. High temperatures adversely affect the productivity of many plant species of which sweet pepper is of no exception.

High humidity in the environment and moist soil together with optimum temperature result in high incidence of various diseases. Diseases like bacterial spot (*Xanthomonas campestris* pv. *vesicatoria*), *Cercospora* leaf spot (*C. capsici*), anthracnose (*Colletotrichum capsici*) and virus diseases are seed borne in nature. To avoid the occurrence of such diseases, seed treatment with various chemicals has been recommended from time to time [7]. But in present day agriculture, use of chemicals for crop production is discouraged. Hence, other alternative treatments for disease control must be developed. One such treatment is with hot water which is economically as well as temporally feasible. Farmers, along with a little techni-cal assistance, can easily adopt this treatment. This treatment is also successful for destroying viruses like *mosaic virus* that affect bell pepper. It is very efficient in destroying pathogens borne both outside the testa and inside the seed testa. In this area no such work has been done in the case of bell pepper.

2. Disease of sweet pepper

Bacterial spot (*Xanthomonas campestris* pv. *vesicatoria*), bacterial wilt (*Pseudomonas solanacearum*), anthracnose/ripe fruit (*Colletotrichum* spp.), *Cercospora* leaf spot/frog eye leaf spot (*C. capsici*), bacterial canker (*Clavibacter michiganensis*) and viruses like Tomato spotted wilt virus (TSWV) are some of the most important diseases of sweet pepper that cause huge economic losses to the farmers. Seed-borne diseases of bell pepper are the most important problems in organic farming systems because of the limitation in chemical control methods.

For the successful management of any disease under normal conditions, sanitation, elimination of primary source and chemical protection at initial stages are some of the measures recommended. Various workers are now engaged in developing and testing the non-chemical methods for seed-borne diseases of vegetables including bell pepper.

3. Seed technology

Seed is the primary and essential starting point of a wide range of horticultural crops, including the majority of vegetables. The high-quality seeds of vegetable varieties that display early, consistent, dynamic seedlings and better- and good-quality fruits from individual seed sown at favourable or unfavourable conditions have increased significantly in recent years. Seedling emergence and field stand establishment is one of the problems faced by the growers, especially in early planting where adverse conditions are prevailing (low temperature and high soil moisture). Delayed, erratic germination and emergence, poor stand, slow early seedling growth rate and non-uniform maturity often limit crop production even under optimum environmental conditions [8–10]. Extensive seed germination and seedling appearance has increased the occurrence of pre-damping off mortality caused by soil-borne fungi [11]. This also leads to establishment of weeds in the fields even before the crop seedlings are mature enough to be cultivated, competing with the main crop for nutrients, and moreover, they hinder the processes of fertilisation, chemical application and mechanical harvesting.

4. Hot water treatment

The most appropriate seed treatment with respect to least damage, economy, efficiency and application is hot water soaking. It is an old-age practice based on treatment with hot water whose temperature is high enough to kill the pathogen but not high enough to harm the seed, hence a very good technique to control many seed-borne diseases [12, 13]. Heat treatment may be applied for agricultural commodities by (1) immersion in hot water, (2) exposure to vapour heat, (3) exposure to hot dry air, (4) treatment with infrared radiation or (5) microwave radiation. Hot water treatments of seed and plant material are classical thermophysical methods of plant protection and are more eco-friendly and effective than chemical treatments.

Hot water treatment can be damaging or not practical for seeds of peas, beans, cucumbers, lettuce, sweet corn, beets and some other crops [12, 14, 15], but it is highly recommended for pepper, eggplant, tomato, cucumber, carrot, spinach, lettuce, celery, cabbage, turnip, radish and other crucifers. It may also severely damage old seeds, and therefore, a small sample of any seed lot over 1 year should be first treated and then tested for germination to determine the amount of injury that may occur. Hot water treatment is recommended for seeds with surface or deep-seated infections. Effective treatment temperature and duration have to be found out for every vegetable crop and the relevant pathogens. The principle is to eliminate the pathogens as far as possible without decreasing germination of seeds. For example, just a 5-min difference in treatment time can lead to diverse differences in the germination rate of cabbage seed.

A number of tests and studies of heat treatment must be undertaken to optimise the time and temperature that are most adaptable to the seeds to be treated and the pathogens to be killed before practical application. Susceptibility to heat damage may differ among different varieties of plant species [16, 17]. The time/tempera-ture combination for a given plant seed depends on many factors interacting with the heat susceptibility of the host, viz., conditions of external layers, dormancy, moisture content, age and vigour [18]. Particularly, it has long been known that the smaller the initial water content of seeds is at the time of heating, the greater the resistance to high temperatures [19]. Two major groups of proteins may be activated by the hot water treatments that induce fruit resistance: heat shock proteins (HSPs) and pathogenesis-related (PR) proteins. HSPs are believed to play a major role in thermotolerance [20–22]. Among the PR proteins, most characterised enzymes chitinases and β-1,3-glucanases hydrolyse polymers of fungal cell walls and are, therefore, thought to be involved in the plant defence mechanism against fungal infection [23, 24].

5. Effect of hot water treatment on different vegetables

Nega et al. [15] stated that even with longer treatment times, hot water treatment with a temperature of 40°C had no significant effect on the seed-borne pathogens. However, on all the crops investigated, hot water treatments at temperatures

50 or 53°C for 10–30 min had a good phytosanitary effect. In the majority of cases, these treatment conditions did not affect seed germination. Therefore, to reduce the effect of higher temperature like 53°C on germination, comparatively shorter treatment time must be used, especially on sensitive crops like cabbage, etc. The treatment of 50°C for 30 min is optimal against *Phoma lingam* on cabbage. They also observed that on crops like carrot, cabbage and parsley treatments at temperatures 50–53°C for 10–30 min gave a good elimination of *Alternaria* species.

In the past, Walker [25] observed complete abolition of *P. lingam* from cabbage seeds with treatment at 50°C for 30 min. Clayton [26] recorded similar results for 25 min as well as for 30 min at 50°C. Bant and Storey [27] showed good effects of hot water treatment against *S. apiicola* on celery in field trials. In the view of lack of alternatives, this treatment has proved to be an efficient method on celery and parsley seeds against the *Septoria* species [28] and that too with well-known yield increase. Due to the rapid spread of bacterial diseases like those caused by *Xanthomonas* species [29], hot water treatment has emerged as an important method to control seed-borne bacterial diseases because of lack of chemical or other well-established treatments [30, 31]. The documented efficiency of this treatment on cabbage and cauliflower against *X. campestris* pv. *campestris* is between 25 and 90%. The treatments differ between 50°C for 10–60 min [32–34] and 52°C for 30 min [35].

Melanie et al. [36] demonstrated the efficiency of hot water treatment method in reducing bacterial diseases like bacterial spot and bacterial canker in tomato under greenhouse as well as open-field conditions. They also observed that after treating, seedlings from the same seed lots with hot water did not get diseased in the greenhouse or fields. In plots/fields established from hot water-treated seed, the occurrence of bacterial canker was less extensive, and yields were higher than the plots/fields established from non-treated seeds. Also, fruits from non-treated seeds were considerably smaller than fruits from treated seeds. Reduced infection frequency of bacteria responsible for bacterial canker and bacterial leaf spot was observed in tomato seeds after hot water treatment with increased fruit size and yield.

Hot water treatment of seeds of okra (*Abelmoschus esculentus*) at 52°C for 30 min resulted in the improved crop, both in greenhouse and field conditions. The improvement was with respect to increase in fruit number, leaf number, fruit length and girth, total number of seeds per fruit, seed weight and plant biomass. The hot water treatment of seeds also reduced the frequency of mycoflora infection in the seeds, hence enhancing the vigour index and germination percentage of the seedlings [37]. After soaking carrot seeds in hot water at 52°C for 25 min, the bacterial pathogen (*Xanthomonas hortorum* pv. *carotae*) in and on the carrot was killed [38]. On the other hand, Nandini and Shripad [39], observed that hot water treatment at 52°C for 10 min was effective in controlling the bacterial blight of cowpea with minimum number of infected seedlings and percentage of seedling infection. Germination was also not much affected as compared to control (58 vs. 76.00%).

The effects of hot water treatments of carrot seeds on seed-borne fungi, germination, emergence and yield were studied by Hermansen et al. [40] where the seeds infected with *Alternaria dauci* were hot water treated at temperature ranging from 44 to 59°C at intervals of 5°C for periods of 5–40 min. Treatment of carrot seeds with hot water at 44, 49 and 54°C improved germination rate and reduced the occurrence of *A. dauci*. Hot water treatment at 54°C for 20 min inhibited *A. dauci* without negatively affecting germination rate or yield. Ranganna et al. [41] demon-strated that the potato tubers can safely be stored without sprouting for 12 weeks at 8 or 18°C, if treated with 57.5°C hot water for 20–30 min.

Hot water treatment of seeds was also observed to be helpful in controlling seed-borne pathogens in sweet pepper. Aguilar et al. [42] observed that hot water treatment of bell pepper at 45°C for 15 min or 53°C for 4 min before storing them at 8°C reduced the occurrence of fungal infections. Several hot water treatments of bell pepper seeds resulted in considerable drop-off in seed viability but had no effect on seed vigour [43]. No study can be found in the literature that attempted to arrive at the optimum time-temperature combination for sweet pepper. Therefore, the effect of hot water treatments of sweet pepper seeds on seed viability and seedling vigour needs to be investigated (**Table 1**).

Causal agent	Host	Thermotherapy tested	Result	Ref.
Clavibacter michiganensis subsp. *michiganensis*	Tomato (*Lycopersicon esculentum*)	Hot water treatment at 53, 54 and 55°C for 10–60 min	Germination remained unaffected up to 55°C for 30 min; bacteria were recovered after 30 min at 53 and 54°C and not after 40 min	Bryan [44]
C. m. subsp. *michiganensis*	Tomato	Soaking infected seeds 30 min in hot water at 56°C	Plants develop significantly less disease than when no seed treatment is used; seed germination is slightly reduced	Shoemaker and Echandi [45]
P. s. pv. *phaseolicola*	Bean (*Phaseolus vulgaris*)	Hot water treatments: 50°C for 45–60 min	Reduction of bacterial number by 98–100% but 45% reduction in seed germination for 60 min soaking with naturally infected seeds	Tamietti [46]; Tamietti and Garibaldi [47]
P. s. pv. *pisi*	Pea (*Pisum sativum*)	Dry heat at 65°C for 72 h and soaking in water at 55°C for 15 min	Significant reduction of pathogen contamination but germination lowered 5–20%	Grondeau et al. [48]
P. s. pv. *tomato*	Tomato	Infected seeds are subjected to hot water treatment at 48°C for 60 min	Pathogens are killed, and germination is not affected	Devash et al. [49]
Xanthomonas campestris pv. *campestris*	Cabbage, cauliflower (*Brassica oleracea*)	Hot water treatment at 50°C for 30 min	Pathogens are eliminated successfully by this treatment to prevent seedling infestation	Shekhawat et al. [34]
X. c. pv. *carotae*	Carrot (*Daucus carota*)	Hot water treatment (52°C for 10 min)	Can prevent pathogen infestation	Ark and Gardner [50]
X. c. pv. *cucurbitae*	Pumpkin (*Cucumis pepo*)	Hot water treatment at 54 and 56°C for 30 min	Greatly reduces the level of seed infestation but does not completely eliminate it	Moffett and Wood [51]

Adopted from Grondeau et al. [52].

Table 1.
Thermotherapy to free seeds from pathogenic bacteria.

6. Conclusion

High incidences of disease in consumers leading to fatal diseases like cancer have attracted the attention of researchers. The causative agents were explored, and the focus was indiscriminate use of chemicals starting from seed treatment to crop productions. The residual effects of pesticides, herbicides and other chemicals are long lasting, adversely affecting the health of consumers. Some mechanism and treatments are needed to address these problems. Hot water treatment is one such mechanism which has been reviewed in the chapter to help the consumers taking care of all the public health issues. The present study provides easy to practice technique to farmers without involving cumbersome techniques to farmers to protect the capsicum against bacterial, viral and fungal diseases, leading to optimal productions without harming the interest of consumers.

Author details

Suryapal Singh[1*], Harshita Singh[2] and Narender K. Bharat[3]

1 Department of Seed Science and Technology, Chaudhary Charan Singh Haryana Agricultural University, Hisar, Haryana, India

2 Department of Vegetable Science, Chaudhary Charan Singh Haryana Agricultural University, Hisar, Haryana, India

3 Department of Seed Science and Technology, Dr. Y.S. Parmar University of Horticulture and Forestry, Nauni Solan, HP, India

*Address all correspondence to: surajahlawat06@gmail.com

References

[1] Manchanda AK, Singh B. Effect of plant density and nitrogen on yield and quality of bell pepper (*Capsicum annum* L). Indian Journal of Horticulture. 1987;**44**:250-252

[2] Agarwal A, Gupta S, Ahmed Z. Influence of plant densities on productivity of bell pepper (*Capsicum annuum* L.) under greenhouse in high altitude cold desert of Ladakh. Acta Horticulture. 2007;**756**:309-314

[3] Sun T, Xu Z, Wu CT, Janes M, Prinyawiwatkul W, No HK. Antioxidant activity of different colored sweet bell peppers (*Capsicum annuum* L.). Journal of Food Science. 2007;**72**:98-102

[4] Singh OP, Anand N, Deshpandey AH. Improvement of bell pepper. Advances in Horticulture. 1993;**5**:87-104

[5] NHB. Final Area and Production Estimates for Horticultural Crops for 2013-14. Gurgaon, India: National Horticultural Board; 2015. Available from: http://www.nhb.gov.in/area%20_production

[6] Bosland PW, Votava EJ. Peppers: Vegetables and Spice Capsicum. New York: CABI Publishing; 1999

[7] Gupta SK, Thind TS. Disease Problem in Vegetable Production. Jodhpur, India: Scientific Publisher; 2006. pp. 342-346

[8] Cavero J, Ortega RG, Zaragoza C. Influence of fruit ripeness at the time of seed extraction on peppers (*Capsicum annuum* L.) seed germination. Scienta Horticulturae. 1995;**60**:345-352

[9] Gerson R, Honma S. Emergence response of the pepper at low soil temperature. Euphytica. 1978;**27**:151-156

[10] McGrady JJ, Cotter DJ. Anticrustant effects on Chile pepper stand and yield. Horticulture Science. 1984;**19**:408-409

[11] Ferriss RS, Baker JM. Relationships between soybean seed quality and performance in soil. Seed Science and Technology. 1990;**18**:51-73

[12] Floyd R. Vegetable Seed Treatments, Farm Note 90/190. Western Australia: Department of Agriculture and Food; 2005

[13] Muniz MFB. Control of microorganisms associated with tomato seeds using thermotherapy. Revista Brassileira Sementes. 2001;**23**(1):176-280

[14] Miller SA and Ivey LLM. Hot Water Treatment of Vegetable Seeds to Eradicate Bacterial Plant Pathogens in Organic Production Systems The Ohio State University Extension, Columbus, OH Ohio State Extension Bulletin HYG-3086-05; 2005. pp. 3

[15] Nega E, Ulrich R, Werner S, Jahn M. Hot water treatment of vegetable seed—An alternative seed treatment method to control seed-borne pathogens in organic farming. Journal of Plant Disease and Protection. 2003;**110**:220-234

[16] Lozoya-Saldana H, Merlin-Lara O. Thermotherapy and tissue culture for elimination of Potato Virus X (PVX) in Mexican potato cultivars resistant to late blight. American Potato Journal. 1984;**61**(12):735-739

[17] Smoot JJ, Segall RH. Hot water as a postharvest control of mango anthracnose. Plant Disease Report. 1963;**47**(8):738-742

[18] Baker KF. Thermotherapy of planting material. Phytopathology. 1962;**52**:1244-1255

[19] Waggoner HD. The viability of radish seeds(*Raphanus sativus* L) as affected by high temperatures and water

content. American Journal of Botany. 1917;**4**:299-313

[20] Howarth CJ, Ougham HJ. Gene expression under temperature stresses. New Phytology. 1993;**125**:1-26

[21] Sabehat A, Weiss D, Lurie S. Heat-shock proteins and cross-tolerance in plants. Physiologia Plantarum. 1998;**103**:437-441

[22] Vierling E. The roles of heat shock proteins in plants. Annual Review of Plant Physiology and Plant Molecular Biology. 1991;**42**:579-620

[23] Collinge B, Kragh KM, Mikkelsen JD, Nielsen KK, Rasmussen U, Vad K. Plant chitinases. Plant Journal. 1993;**3**:31-40

[24] Schlumbaum A, Mauch F, Vogeli U, Boller T. Plant chitinases are potent inhibitors of fungal growth. Nature. 1986;**324**:365-367

[25] Walker JC. The hot water treatment of cabbage seed. Phytopatholoy. 1923;**13**:251-253

[26] Clayton EE. Seed treatment for black-leg diseases of crucifers. New York Agricultural Experiment Station (Geneva) Technical Bulletin. 1928:137

[27] Bant JH, Storey IF. Hot water treatment of celery seed in Lancashire. Plant Pathology. 1952;**1**:81-83

[28] Merz F. Pflanzenschutztabellen für den Erwerbsgemüsebau. Gemüse. 2000;**36**:1-39

[29] Poschenrieder P. Pflanzliche Bakteriosen in Bayern– Beobachtungen, Untersuchungen und Erfahrungen in den achtziger und neunziger Jahren. Gesunde Pflanzen. 2000;**52**:212-220

[30] Koch I, Wahl R, Laun N, Krauthausen HJ, Bauermann W. Bakterielle Blattflecken an Möhren - Bedeutung und Gegenmaßnahmen. Gemüse. 2000;**36**:28-30

[31] Linders R. Aktueller Wissensstand bei *Xanthomonas campestris* an Kohl. Gemüse. 2000;**36**:31-32

[32] Babadost M, Deric ML, Gabrielson RL. Efficacy of sodium hypochlorite treatments for control of *Xanthomonas campestris* pv. *campestris* in Brassica seeds. Seed Science and Technolology. 2006;**24**:7-15

[33] Sha A, Srivastava K, Roy AJ, Bora SS. Control of black rot disease of cauliflower by seed treatment. Progressive Horticulture. 1985;**17**:72-74

[34] Shekhawat PS, Jain ML, Chkravarti BP. Detection and seed transmission of *Xanthomonas campestris* pv. *campestris* causing black rot of cabbage and cauliflower and its control by seed treatment. Indian Phytopathology. 1982;**35**(3):442-447

[35] Sharma SL. Control of black rot of cauliflower by hot water seed treatment and field sprays with streptocycline. Indian Journal of Mycology and Plant Pathology. 1980;**11**:17-20

[36] Melanie L, Ivey L, Miller SA. Evaluation of hot water seed treatment for the control of bacterial leaf spot and bacterial canker on fresh market and processing tomatoes. Acta Horticulture. 2005;**695**:198-200

[37] Begum M, Lokesh S. Effect of hot water and ultra violet radiation on the incidence of seedborne fungi of okra. Archives of Phytopathology and Plant Protection. 2012;**45**:126-132

[38] Temple TN, Toit LJ, Derie ML, Johnson KB. Quantitative molecular detection of *Xanthomonas hortorum* pv. *carotae* in carrot seed before and after hot-water treatment. Plant Disease. 2013;**97**:1585-1592

[39] Nandini R, Shripad K. Evaluation of hot water treatment on seed germination and seedling infection of artificially inoculated cowpea seeds by *Xanthomonas axonopodis* pv. *vignicola*. International Journal of Bioassays. 2015;**4**(8):4181-4183

[40] Hermansen A, Brodal G, Balvoll G. Hot water treatments of carrot seeds: Effects on seed-borne fungi, germination, emergence and yield. Seed Science and Technology. 1999;**27**(2):599-613

[41] Ranganna B, Raghavan GSV, Kushalappa AC. Hot water dipping to enhance storability of potatoes. Postharvest Biology and Technology. 1998;**13**(3):215-223

[42] Aguilar GA, Cruz R, Baez R. Storage quality of bell peppers pretreated with hot water and polyethylene packaging. Journal of Food Quality. 1998;**22**:287-299

[43] Musazura W, Bertling I. Investigations into the effect of multiple hot water treatment of tomato (*Solanum lycopersicum*) and pepper (*Capsicum annuum*) seeds on seed viability and seed vigor. Acta Horticulturae 2013;**1007**:795-799

[44] Bryan ML. Studies on bacterial canker of tomato. Journal of Agricultural Research. 1930;**41**(12):825-851

[45] Shoemaker PB, Echandi E. Seed and plant bed treatments for bacterial canker of tomato. Plant Disease Report. 1976;**60**(2):163-166

[46] Tamietti G. Seed treatment trials to control halo blight of bean. Informatore Fitopatologico. 1982;**32**(6):47-50

[47] Tamietti G, Garibaldi A. Eradication of *Pseudomonas syringae* pv. *phaseolicola* in bean seeds in Italy. In: Quacquarelli A, Casano FJ, editors. Proc. Works. "Phytobacteriology and Plant Bacterial Diseases of Quarantine Significance," Roma, Italy, 17-21 September 1984. Roma, Italy: Instituto Sperimentale per la PatologiaVegetale; 1984

[48] Grondeau C, Ladonne F, Fourmond A, Poutier F, Samson R. Attempt to eradicate *Pseudomonas syringae* pv. *pisi* from pea seeds by heat treatments. Seed Science and Technology. 1992;**20**:515-525

[49] Devash Y, Bashan Y, Okon Y, Henis Y. Survival of Pseudomonas tomato in soil and seeds. Hassad. 1979;**60**(3):597-601

[50] Ark PA, Gardner MW. Carrot bacterial blight as it affects the roots. Phytopathology. 1944;**34**:416-420

[51] Moffett ML, Wood BA. Seed treatment for bacterial spot of pumpkin. Plant Disease Reporter. 1979;**63**(7):537-539

[52] Grondeau C, Samson R, Sands DC. A review of thermotherapy to free plant materials from pathogens, especially seeds from bacteria. Critical Reviews in Plant Sciences. 1994;**13**(1):57-75

6

Hot Pepper (*Capsicum annuum* L.): An Alternative Food to Reduce Micronutrient Deficiencies in Human

S.R. Krishna Motukuri and Nallamothu Jaswanthi

Abstract

Hot peppers are good source of bioactive compounds particularly phytochemicals, capsaicin, oleoresins, and rich organic micronutrients with potential health benefits. It plays an important role in diets and possesses micronutrients. Micronutrient dietary deficiency remains a massive problem in the world which may cause several chronic health diseases. More than 2 billion individuals are facing micronutrient deficien-cies, viz., zinc, iodine, and iron, followed by vitamins. Among various approaches to overcome human nutrition deficiencies, a diet with fruits and vegetables that are rich in micronutrients is considered as the best solution. Hot pepper consump-tion worldwide is well known for its high nutritional content which in turn reduces human micronutrient deficiencies. Thus sufficient amounts of micronutrients can be provided by incorporating the nutrient-rich pepper in diet that could beneficially help in combating nutrient deficiencies. Thus, in the present chapter, an overview of the micronutrient deficiency is described, and the nutrition status of hot pepper that utilized in eradicating human micronutrient dietary deficiencies is also discussed.

Keywords: *Capsicum*, hot pepper, micronutrients, human dietary deficiency, advanced approaches

1. Introduction

Micronutrients (minerals and vitamins) are obtained from our diet; though required in small amounts, they are necessary for good growth and development [1]. Micronutrients, namely, iron, cobalt, manganese, boron, molybdenum, chlorine, nickel, zinc, copper, iodine, fluorine, and chromium, are essential for human health. Many studies showed that nutrition plays a vital role in providing the substances essential for the formation of the early brain structure and also sup-porting for its good functioning [2]. Micronutrients play a unique metabolic role as cofactors in the absorption and digestion of carbohydrates, proteins, amino acids, and lipids to release energy [3]. Micronutrients play a very important role in all the metabolic activities like cell signaling, motility, cell division, and cell differentiation and death and, in turn, regulate the tissue development [4].

Micronutrient deficiencies can badly affect the health of individuals, which influence the world's morbidity and mortality [5]. These deficiencies are referred to as "hidden hunger" which affects about 2 billion people [6] and may lead to high rates of illness, obesity, underdevelopment, and even deaths in children of age below 5 years [7]. Among various vegetables and fruits that are rich in micronutrients, chilies have various essential nutrients, minerals, vitamins (A, C, E), and other important phytochemical compounds [8]. They also pose some human health benefits like preventing diseases such as obesity, heart diseases, and different cancers [9]. An alkaloid, namely, capsaicin, is present in pepper which has antimicrobial, anti-inflammatory, and anticancer effects on the digestive system and is used in relieving pain and also to lose weight [10]. Thus providing diet that is rich in pepper can be supportive in an ongoing quest to ease micronutrient deficiencies.

2. Micronutrient deficiency in humans

Some studies reveal the importance of micronutrients in pregnant women and generally given as supplements, which include vitamins, viz., A, cobalamin, pyridoxine, C, D, and E, iron, zinc, iodine, copper, and selenium. Micronutrient deficiencies may mimic radiation or some chemicals which damage the nucleic acids and lead to cancer [11]. Some studies reveal that these deficiencies are related with the danger of HIV infection progression and deaths [12]. Deficiency of vitamin A may cause low serum level of retinol which affects nearly 15% of pregnant women and which leads to night blindness in 8% of them [4]. The reduced form of vitamin C [ascorbic acid (AA)] is the principal chemical structure that appears as an important micronutrient that involves in various physiological functions. Ascorbic acid plays a vital role in reducing the antioxidants and enhances the production of ROS to prevent cancer [13]. Vitamin D is generally known as fat-soluble compound with antiproliferative effect and also involved in the development of bone and immune system [14]. Some studies reported that vitamin D metabolites help in protection against cancer [15].

Vitamin E deficiency can lead to enhanced peroxidation which leads to symptoms like walking difficulties and severe development of speech, reduced fat absorption syndrome, and lipoprotein abnormalities [16]. Deficiencies of iron may affect the growth and mental development and also decreases the capability to do physical work [3]. Some studies reported that the supplementation of zinc along with other micronutrients can reduce the severity of diarrhea diseases and respiratory infections [17]. These deficiencies occur in individuals who do not consume food that provides micronutrients sufficiently like fortified foods, fruits, and animal products. This is usually due to its excessive cost or may domestically unreachable. Among the various vegetables available, chili (*Capsicum annuum*), which are available to common man at affordable price and provide various vitamins like E, C, B6, B12, and provitamin A and some minerals, possibly will give rise to significantly enhanced nutrition [8].

3. Role of hot pepper in micronutrient deficiencies

To enhance the micronutrient status in humans, nutritionists suggest integrating foods, which are rich in micronutrients, in diet. Among various vegetables, chili fruits are rich in capsaicinoids, carotenoids, tocopherols, provitamin A, ascorbic acid, and several antioxidants. Chilies are also an excellent source of xanthophylls and vitamins B1 (thiamine), B3 (niacin), and P (citrin) [18]. Hot pepper contains proteins (1.9 g), sugars (5.3 g), fiber (1.5 g), fat (0.4 g), energy (8.8 g), ascorbic acid (240%), pyridoxine (39%), vitamin A (32%), copper (14%), iron (13%), potas-sium (7%), and magnesium (6%) [19]. Nutrient constituents and their composition present in chili species in different countries are represented in **Table 1**. Red sweet

So. no	Constituent	Name of the country	Fruit species	Composition	References
1	Vitamin C	United States	*C. Francisca*	122.0 mg/100 g DW	[20]
		Brazil	*C. chinense*	125 mg/100 g DW	[21]
		India	*Capsicum chinense Jacquin*	109.36 mg/100 g DW	[22]
		Ethiopia	*C. annuum*	84.011–89.011 mg/100 g DW	[23]
2	Capsaicin	Brazil	*C. chinense*	14.0 mg/g DW	[24]
		Central America	*C. annuum*	0.042 mg/g DW	[25]
		India	*C. frutescens*	4.45 mg/g DW	[26]
		Ethiopia	*C. annuum (Marako fana)*	5.5 mg/g DW	[27]
3	Crude protein	America	*C. annuum*	2.9 g/100 g DW	[28]
		India	*C. chinense*	17.5 g/100 g DW	[22]
		Ethiopia	*C. annum (Marako fana)*	118.09 g/100 g DW	[29]
		Brazil	*C. chinense*	0.08–4.5 g/100 g DW	[21]
4	Carbohydrates	America	*C. annuum*	3.0 g/100 g DW	[28]
		India	*C. chinense*	78.1 g/100 g DW	[22]
		Brazil	*C. chinense*	1.8–10.8 g/100 g DW	[21]
		Ethiopia	*C. annum (Marako fana)*	35.3 ± 0.6 g/100 g DW	[29]

Table 1.
Nutrient constituents and their composition in chili species in different countries.

pepper consists of twice the provitamin A than in carrot and double the amount of vitamin C of green pepper, and it also acts as antioxidant and anti-inflammatory phytonutrient [30]. Sweet pepper is fibrous in nature and is the sources of dietary fiber which could help in fighting blood cholesterol level and diabetes better from the body and bloodstream, respectively [31].

4. Medicinal uses of hot pepper

Capsaicin is a pungent alkaloid present in hot pepper species, and it is the principal capsaicinoid that accounts for about 71% of the pungent types, followed by dihydrocapsaicin [32]. Structures of capsaicin and dihydrocapsaicin are in **Figure 1**. The percentage of pungency varies among species by varying the capsaicin and dihydrocapsaicin contents which are due to influence of factors like the developmental stages of fruit and the environmental conditions [33]. Capsaicin possesses good medicinal properties like antimicrobial, anticancer, antidiabetic, and analgesic properties. Some studies reveal that fruits having higher amounts

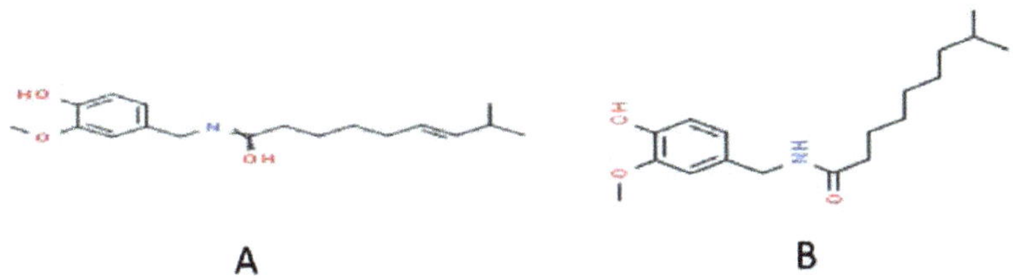

Figure 1.
(A) Capsaicin. (B) Dihydrocapsaicin.

of capsaicin show high antioxidant level [32]. Besides the nutritional benefits, it is also responsible for medical therapeutic and pharmacological uses [34]. Capsaicin has an important role in relieving sore throats, fever, and cold symptoms and also enhances the circulation of blood and strengthens the arteries, thus reducing the risk of heart attacks [3].

Capsaicin is also used as an anti-irritant balm for external use and as painkiller in some topical ointments, nasal sprays, and skin patches and also used in the form of cream for short-term relaxation of muscles and pains related to arthritis, back pain, and other stresses [35]. It also has an antidiabetic activity by enhancing the insulin secretion and releasing bound insulin. Capsaicin helps in producing heat within the body and consumes oxygen which simultaneously burns calories in the body and helps in losing weight [30]. Thus, incorporating pepper-rich food in everday diet can be useful in enduring pursuit of eliminating micronutrient deficiency. Hot pepper fruits were enriched with antioxidants which protect the human body from harmful effects of free radicals, and it develops resistance against different diseases.

5. Conclusion

Advanced technologies are being used to combat human micronutrient deficiency in developing countries, where poor people are mostly at this risk. In tropical countries, rural farmers traditionally cultivate wide range of vegetables; among them chili is grown globally, and India is the most important producer and exporter of hot pepper which consists of wide range of micro- and macronutrients. In this chapter, it has been concluded that the chilies are cost-effective vegetables that are significantly rich in essential micro- and macronutrients and also possess healthy components to support human health conditions. The presence of antioxidants and phytochemicals in chili increases its importance in controlling the diseases. Integrating hot pepper in our regular diet can be supportive to alleviate nutrient deficiency in humans.

Acknowledgements

The authors thank the head of the department, Department of Biotechnology, Koneru Lakshmaiah Education Foundation deemed to be University, Guntur, Andhra Pradesh, India, for providing support in writing this book chapter.

Author details

S.R. Krishna Motukuri* and Nallamothu Jaswanthi
Department of Biotechnology, Koneru Lakshmaiah Education Foundation, Guntur, Andhra Pradesh, India

*Address all correspondence to: msrkrishna81@gmail.com

References

[1] McMillan DC, Maguire D, Talwar D. Relationship between nutritional status and the systemic inflammatory response: Micronutrients. Proceedings of the Nutrition Society. 2019;**78**:56-67. DOI: 10.1017/S0029665118002501

[2] Prado EL, Dewey KG. Nutrition and brain development in early life. Nutrition Reviews. 2014;**72**:267-284. DOI: 10.1111/nure.12102

[3] Olatunji Tomi L, Anthony J. Afolayan. The suitability of chili pepper (*Capsicum annuum* L.) for alleviating human micronutrient dietary deficiencies: A review. Food Science & Nutrition. 2018;**6**:2239-2251. DOI: 10.1002/fsn3.790

[4] Gernand AD, Schulze KJ, Stewart CP, West KP, Christian P. Micronutrient deficiencies in pregnancy worldwide: Health effects and prevention. Nature Reviews. Endocrinology. 2016;**12**:274-289. DOI: 10.1038/nrendo.2016.37

[5] Shivakoti R, Christian P, Yang WT, Gupte N, Mwelase N, Kanyama C, et al. Prevalence and risk factors of micronutrient deficiencies pre- and post-antiretroviral therapy (ART) among a diverse multicountry cohort of HIV-infected adults. Clinical Nutrition. 2016;**35**:183-189. DOI: 10.1016/j.clnu.2015.02.002

[6] Bailey RL, West KP, Black RE. The epidemiology of global micronutrient deficiencies. Annals of Nutrition & Metabolism. 2015;**66**:22-33. DOI: 10.1159/000371618

[7] Global Nutrition Report. Global Nutrition Report 2017: Nourishing the SDGs (Development Initiatives, 2017)

[8] Materska M, Perucka I. Antioxidant Activity of the Main Phenolic Compounds Isolated from Hot Pepper Fruit (*Capsicum annuum* L.). Journal of Agricultural and Food Chemistry. 2005;**53**:1750-1756. DOI: 10.1021/jf035331k

[9] Alzrog AM, Mohamed AS, Zakaria RB, Alias AKB. Effect of planting media (rice husk and coco peat) on the uptake of cadmium and some micronutrients in chilli (*Capsicum annum* l.). Pure and Applied Biology. 2013;**2**:76. DOI: 10.19045/bspab.2013.23001

[10] Srinivasan K. Biological activities of red pepper (*Capsicum annuum*) and its pungent principle capsaicin: A review. Critical Reviews in Food Science and Nutrition. 2015;**56**:1488-1500. DOI: 10.1080/10408398.2013.772090

[11] Ames BN. DNA damage from micronutrient deficiencies is likely to be a major cause of cancer. Mutation Research, Fundamental and Molecular Mechanisms of Mutagenesis. 2001;**475**:7-20. DOI: 10.1016/S0027-5107(01)00070-7

[12] Drain PK, Kupka R, Mugusi F, Fawzi WW. Micronutrients in HIV-positive persons receiving highly active antiretroviral therapy. The American Journal of Clinical Nutrition. 2007;**85**:333-345. DOI: 10.1093/ajcn/85.2.333

[13] Heaney ML, Gardner JR, Karasavvas N, Golde DW, Scheinberg DA, Smith EA, et al. Vitamin C antagonizes the cytotoxic effects of antineoplastic drugs. Cancer Research. 2008;**68**:8031-8038. DOI: 10.1158/0008-5472.CAN-08-1490

[14] Gandini S, Raimondi S, Gnagnarella P, Dore JF, Maisonneuve P, Testori A. Vitamin D and skin cancer: A meta-analysis. European Journal of Cancer. 2009;**45**:634-641. DOI: 10.1016/j.ejca.2008.10.003

[15] Deeb KK, Trump DL, Johnson CS. Vitamin D signalling pathways in cancer: Potential for anticancer therapeutics. Nature Reviews Cancer. 2007;7:684-700. DOI: 10.1038/nrc2196

[16] Mamede AC, Tavares SD, Abrantes AM, Trindade J, Maia JM, Botelho MF. The role of vitamins in Cancer: A review. Nutrition and Cancer. 2011;**63**:479-494. DOI: 10.1080/01635581.2011.539315

[17] Bhutta ZA. Iron and zinc deficiency in children in developing countries. British Medical Journal. 2007;**334**:104-105. DOI: 10.1136/bmj.39094.513924.BE

[18] Bosland PW, Votava EJ. Peppers: Vegetable and Spice Capsicums. Wallingford, CT: CABI Publishing; 2000. p. 204

[19] Ebong PE. Quality and Safety of Medicinal Plants in Phytomedicine: Panacea for Primary Health Care Delivery in Nigeria. Nigeria: University of Calabar Press; 2015. pp. 293-310

[20] Teodoro AF, Alves RD, Ribeiro LB, Reis K, Reifschneider FJ, Fonseca ME, et al. Vitamin C content in Habanero pepper accessions (Capsicum chinense). Horticultura Brasileira. 2013;**31**:59-62. DOI: 10.1590/S0102-05362013000100009

[21] Da Silveira A-CT, da Silva GI, de Melo LAMP, Reifschneider FJB, da Costa Ribeiro CS. Carotenoid and total vitamin C content of peppers from selected Brazilian cultivars. Journal of Food Composition and Analysis. 2017;**57**:73-79. DOI: 10.1016/j.jfca.2016.12.020

[22] Malakar S, Sarkar S, Kumar N, Jaganmohan R. Studies of biochemical characteristics and identification of active phyto-compounds of king chili (Capsicum chinense Jacq.) using GC-MS. Journal of Pharmacognosy and Phytochemistry. 2018;7:3100-3104

[23] Kinfe E. Nutritional composition, physicochemical and functional properties of some capsicum varieties grown in Ethiopia [Doctoral dissertation]. Addis Ababa University. 2009. Available from: //localhost:80/xmlui/handle/123456789/8644

[24] Rêgo ER, Finger FL, Rêgo MM. Types, uses and fruit quality of Brazilian chili peppers. In: Kralis JF, editor. Spices: Types, Uses and Health Benefits. Nova Science Publishers; 2011. 70p

[25] Cázares-Sánchez E, Ramírez-Vallejo P, Castillo-González F, Soto-Hernández RM, Rodríguez-González MT, Chávez-Servia JL. Capsaicinoids and preference of use in different morphotypes of chili peppers (*Capsicum annuum* L.) of east-Central Yucatán. Agrociencia. 2005;**1**:627-638

[26] Tilahun S, Paramaguru P, Rajamani K. Capsaicin and ascorbic acid variability in chilli and paprika cultivars as revealed by HPLC analysis. Journal of Plant Breeding and Genetics. 2013;**1**:85-89

[27] Aklilu Alemu SH. Diversity, disease reaction, heterosis and combining ability in hot pepper (*Capsicum annum* L.) genotypes for agro-morphological and quality traits in ethiopia [Doctoral dissertation]. Asela University. 2018. Available from: //hdl.handle.net/123456789/2253

[28] Kumar S, Kumar R, Singh J. Cayenne/American pepper. In: Handbook of Herbs and Spices. Vol. 3. Cambridge, UK: Woodhead Publishing; 2006. pp. 299-312. DOI: 10.1533/9781845691717.3.299

[29] Esayas K, Shimelis A, Ashebir F, Negussie R, Tilahun B, Gulelat D. Proximate composition, mineral content and antinutritional factors of some capsicum (Capsicum annum) varieties grown in Ethiopia. Bulletin of the Chemical Society of Ethiopia. 2011;**25**. DOI: 10.4314/bcse.v25i3.68602

[30] Mercy ER, David U. Potential health benefits of conventional nutrients and phytochemicals of Capsicum peppers. Pharmacy & Pharmacology International Journal. 2018;**6**:62-69. DOI: 10.15406/ppij.2018.06.00157

[31] Mateljan G. The World's Healthiest Foods. USA: The George Mateljan foundation; 2018

[32] Chu CJ, Huang SM, de Petrocellis L, Bisogno T, Ewing SA, Miller JD, et al. N-Oleoyldopamine, a novel endogenous capsaicin- like lipid that produces hyperalgesia. The Journal of Biological Chemistry. 2003;**278**:13633-13639. DOI: 10.1074/jbc.M211231200

[33] Estrada B, Bernal MA, Díaz J, Pomar F, Merino F. Capsaicinoids in vegetative organs of Capsicum annuum L. in relation to fruiting. Journal of Agricultural and Food Chemistry. 2002;**50**:1188-1191. DOI: 10.1021/jf011270j

[34] Peña-Alvarez A, Ramírez-Maya E, Alvarado-Suárez LA. Analysis of capsaicin and dihydrocapsaicin in peppers and pepper sauces by solid phase microextraction – Gas chromatography – Mass spectrometry. Journal of Chromatography. A. 2009;**1216**:2843-2847. DOI: 10.1016/j.chroma.2008.10.053

[35] Fattori V, Hohmann MS, Rossaneis AC, Pinho-Ribeiro FA, Verri WA. Capsaicin: Current understanding of its mechanisms and therapy of pain and other pre-clinical and clinical uses. Molecules. 2016;**21**:844. DOI: 10.3390/molecules21070844

7

Anticancer Effect of Capsaicin and its Analogues

Balasubramanian Arul and Ramalingam Kothai

Abstract

Potent biomolecules from natural products from plants, animals, and minerals are the fundamental basis of the ailment of mankind. *Capsicum* or red pepper plants were grouped under the kingdom Plantae and family Solanaceae. It is used widely throughout the world in foods for their pungent flavor and aroma, and to prolong food spoilage. This chapter presents a frame of a concise compilation of the anticancer and cytotoxic potentials of *Capsicum*, its analogs, and related compounds. Capsaicin (trans-8-methyl-N-vanillyl-6-nonenamide) is the most predominant and naturally occurring alkaloid from the *Capsicum* species. It also details the anticancer efficacy of capsaicin and its analogs like capsaicinoids and capsiates which possess antioxidants and targets multiple signaling pathways, ontogenesis, and tumor-suppressor genes in various types of cancer models. Capsaicin is a major ingredient and has been linked to suppression of growth in various cancer cells. The data available strongly indicate the significant anticancer benefits of capsaicin and its potent analog molecules. It shows a significant effect on cancer cell proliferation, apoptosis, cancer cell surveillance, growth arrest, and metastasis. This chapter also predominantly focuses on the combinational use of capsaicin with other natural dietary compounds as a measure of synergistic anticancer activities.

Keywords: *Capsicum*, capsaicin, apoptosis, angiogenesis, metastasis, anticancer

1. Introduction

Capsicum is one of the most important genera in the Solanaceae family and consists of almost 30 species [1]. Their fruits were commonly referred to as peppers or chili peppers, having a bell shape that appears in different colors such as red, green, orange, and yellow. These have been commonly used in their diet by human society from ancient times as herb and spice. *Capsicum* is widely used as an essential ingredient in kitchens worldwide for their pungent flavor and scent and to prevent food contamination. It is known under many names, such as red pepper, chili pepper, paprika, tabasco, aji, cayenne, and tabasco jalapeno, because of its global consumption [2]. The genus *Capsicum* is comprised of various numbers of wild and few domestic species. The species domesticated are *Capsicum annum, C. baccalaureus, C. honey, C. frutescens, C. pudders,* in which *C. chinense* is the fruit with the highest pungency [3]. *Capsicum* has been documented for its wide range of activities, such as pain killers, cardiovascular protection, a cancer chemopreventive, and antioxidant effects. This chapter provides a comprehensive description of the capsaicin and its analogs as well as, more specifically, look into its therapeutic potential in various human cancer and future therapeutic directions of capsaicin.

2. Capsaicinoids

The chili-fruit pungent principles are called capsaicin. Capsaicin and many related compounds are called capsaicinoids. Capsaicin (trans-8-methyl-N-vanillyl-6-nonenamide) is a crystalline, lipophilic, colorless, and odorless alkaloid that is soluble in fat, alcohol, and oil. The major capsaicinoids or the analogs from *Capsicum annum* species are Capsaicin and 6,7-dihydrocapsaicin. The smaller capsaicinoids are nordihydrocapsaicin, homodihydrocapsaicin, and homocapsaicin. Capsaicin and their analogs differ only in acyl group saturation [4, 5]. These alkaloids are produced solely by the genus *Capsicum*, and they are used as a potent analgesic for the treatment of pain and inflammation related to a number of diseases [6, 7].

Although capsaicin traditionally is associated with the analgesic activity, its unpleasant side effects, such as gastric irritation, stomach cramps, and burning sensation, limit capsaicin's applications as a clinically viable drug. This has led to extensive research focusing on the discovery and rational design of capsaicin analogs of the second generation, which have greater bioactivity than capsaicin. Modifications in the acid portion of capsaicin were found to generate analogs with different degrees of pungency. Three capsaicin analogs of two pungent and one with very low pungency were obtained by using multiple lengths of the acyl chain and chemical replacements in the aromatic ring [8]. The chemical structure of capsaicin and its analogs [9, 10] were shown in **Figure 1**.

Capsaicin (**Figure 1(1)**) and its analogs such as capsanthin (**Figure 1(2)**), capsanthin 3′-ester (**Figure 1(3)**), capsanthin 3′,3-diester (**Figure 1(4)**), capsoru-bin (**Figure 1(5)**), capsorubin 3′-ester (**Figure 1(6)**), and capsorubin—3,6-epoxide (**Figure 1(7)**) containing pharmaceutical products have been marketed under the trade name such as Menthacin, Zostrix, and Capzasin-P for topical applications. In 2009, the European Union (EU) and the USA Food and Drug Administration (FDA)

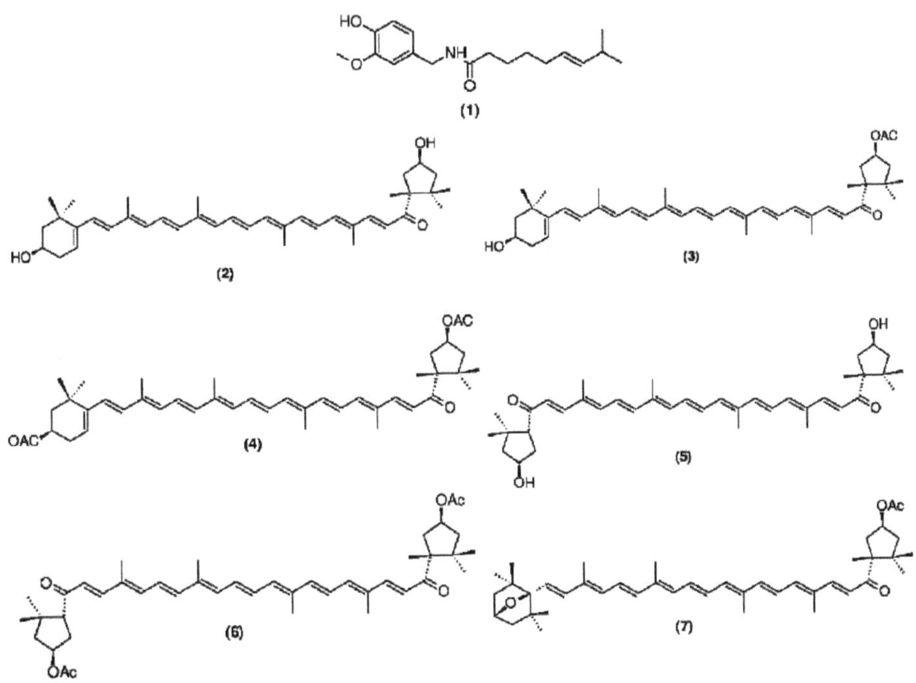

Figure 1.
Capsaicin and its analogs: (1) capsaicin; (2) capsanthin; (3) capsanthin 3′-ester; (4) capsanthin 3′,3-diester; (5) capsorubin; (6) capsorubin 3′-ester; (7) capsorubin—3,6-epoxide.

approved the use of capsaicin 8% patch (Qutenza or NGX-4010) for the treatment of post-herpetic neuralgia (PHN). Resiniferatoxin (RTX) is regarded as an ultra-potent analog of capsaicin that acts as an agonist of Transient receptor potential cation channel subfamily V member 1 (TRPV1) and displays several thousand-fold more potencies than capsaicin [11]. Zucapsaicin, the cis-isomer of capsaicin, shows potent efficacy against episodic cluster migraine prophylaxis, episodic cluster headache, and alleviates neuropathic pain [12]. Thus, capsaicin and its analogs have been used medicinally for centuries, but recently, it has been studied extensively for its analgesic, antioxidant, anti-inflammatory, anti-obesity characteristics and currently for its anticancer activity against a number of types of cancer [13]. It was witnessed by unparalleled advances in the field of capsaicin research. These stud-ies and several other reports clearly showed that capsaicin and its analogs possess multiple pharmacological effects and its application in different clinical conditions.

3. Anticancer activity of capsaicin and its analogs

There is a strong epidemiological and experimental evidence that the phytochemical diet found in fruits, vegetables, whole grains, spices, and teas provides various inhibitory effects against the initiation, development, progression, and metastasis of cancer [14]. Capsaicin, a bioactive phytochemical abundant in chili peppers, is in between them. Capsaicin is a derivative of homovanillic acid, which has been shown to modify the function of many genes associated with cancer cell life span, growth arrest, angiogenesis, and metastasis [15, 16].

Tumorigenesis is a multistage process, which usually begins over an extended period. Cancer cells develop special properties not acquired by most healthy cells. Multiple genetic alterations and aberrant signaling pathways initiate and advance cancer. Determining the molecular targets involved in the tumor development process will provide opportunities to develop a successful cancer-fighting strategy. Studies assessing the capsaicin effect to inhibit cell proliferation by mechanisms are not fully understood in many types of cancer cells [17]. The capsaicin's suggested anticancer pathways include increased cell cycle arrest and apoptosis.

3.1 Capsaicin and apoptosis

Apoptosis is a vital mechanism against the growth of cancer, and it is strongly correlated with the loss of apoptotic signals in malignancy. It has been shown that capsaicin induces apoptosis in many different types of cancer cell lines, including pancreas, colon, prostate, liver, esophagus, bladder, skin, leukemia, lung, and endothelial cells, keeping the normal cells unharmed. A recent review noted capsaicin appears to induce apoptosis in more than 40 distinct lines of cancer cells [13, 18].

Two major signaling systems are the intrinsic mitochondrial death pathway and the extrinsic death receptor pathway, which activate executioner/effector caspases and lead to apoptosis. In specific, the mitochondrial pathway is involved in the complete execution of apoptosis; thus, the mitochondrion has been named as apoptotic mechanism's gatekeeper, and the mitochondrial death pathway proteins and pathways had become important targets for new treatments [19]. Many proteins involved in the mitochondrial death pathway have been targeted by capsaicin to induce apoptosis in different cancer cell lines. For instance, capsaicin treatment activated the cluster of differentiation 95 (CD95)-mediated apoptotic intrinsic and extrinsic pathways [20] and suppressed antiapoptotic protein expression, B-cell lymphoma 2, which causes caspase-9 and -3 activation, loss of mitochondrial membrane potential, and subsequent rises in cytochrome c release [21].

Capsaicin's proapoptotic activity has been found to be mediated via transient vanilloid potential receptor (TRPV1) in many types of cancers [21–23]. It is a nonselective cation channel pertaining to the transient receptor potential channel (TRP) family [24]. It does prefer Ca^{2+} to Na^{+}. Thus it contributes to changes in the concentration of cytosolic free Ca^{2+} and it is capsaicin's primary cell target.

3.2 Reactive oxygen species

Earlier research in pancreatic cells showed that the apoptosis effects of capsaicin were correlated with reactive oxygen species (ROS) production, c-Jun N-terminal Kinase (JNK) activation, mitochondrial depolarization, cytochrome c release in the cytosol, and caspase-3 cascade activation [25]. There is also quite a complex relationship that exists between capsaicin exposure and ROS production. ROS is conventionally considered cytotoxic and mutagenic in normal cells and can induce cell death, apoptosis, and senescence at high levels [26].

Capsaicin has been suggested to induce apoptosis in cancerous cells through the production of higher rates of intracellular ROS. This observation reveals capsaicin as the primary signaling molecule [27, 28]. Capsaicin is capable of activating apoptosis through nonreceptor mechanisms [25].

A number of studies in recent years have shown that oxidative stress causes cellular apoptosis through both mitochondria-dependent and mitochondria-independent pathways [29].

Mitochondria account for 50% of the total cytoplasmic volume in most cells and they participate more than any other organelle in metabolic functions, par-ticularly those involved in cellular energy production. It also consumes nearly 90% of cellular oxygen and is the main source of ROS produced during breathing, and it is engaged in maintaining the intracellular redox state. Along with their long-standing role in energetics, mitochondria depict its prime focus in mammalian cells for many cell death signals. Interactions at the mitochondrion eventually determine whether a cell survives or dies in reaction to many physiological or therapeutic stimuli of cell death. A number of proteins involved in the mitochondrial death pathway were shown to be targeted by capsaicin in order to initiate apoptosis in various cancer cell lines.

It was reported that a 12-h exposure to high concentrations of capsaicin was necessary to induce higher apoptosis rates in Cellosaurus Colo 16 (COLO 16) cells [26]. More than half experienced apoptosis, which was correlated with progressive irreversible dispersion of mitochondrial transmembrane potential and elevated superoxide levels, illustrates the destruction of mitochondria and subsequent breakdown of mitochondrial electron transfer.

Nicotinamide adenine dinucleotide phosphate (NADPH) oxidase is a part of the complex I of the mitochondrial electron transport chain. Capsaicin has been shown to specifically inhibit mitochondrial NADPH oxidase activity by competitively binding this enzyme to the ubiquinone/coenzyme Q site. Therefore, if capsaicin blocks the transportation of electrons in mitochondria, irreversible dispersion of mitochondrial transmembrane potential will occur. This will initiate apoptosis and change the mitochondrial permeability to release cytochrome c and eventual activa-tion of proapoptotic pathways.

Capsaicin can block NAPDH oxidase in the plasma membrane by acting as a Q antagonist coenzyme. Capsaicin's vanillyl moiety is structurally similar to coenzyme Q's cyclic part, which could address the fact that vanilloids serve as antagonists to the coenzyme Q. Suppression of plasma membrane NAPDH oxidase was correlated with capsaicin's prooxidant and proapoptotic properties in some transformed cells, and activated T cells [30].

Another mechanism proposed by capsaicin's anticancer activity is interac-tion with AMP-dependent protein kinase (AMPK), which is the cell's primary metabolic gatekeeper, belonging to the family of protein kinase stimulated during enzyme-depleting metabolic states like hypoxemia, thermal shock, oxidative stress, and physical activity. It acts as a significant metabolic transition for maintaining energy homeostasis and shown to be an intrinsic controller of the mammalian cell cycle [31, 32].

There is an increase in intracellular Adenosine diphosphate (ADP)/Adenosine triphosphate (ATP) and/or Adenosine monophosphate (AMP)/ATP ratios [33] dur-ing energetic imbalance and it promotes activation of AMPK. Activation of AMPK increases oxidative stress in many human cancer cells and induces apoptosis. This stimulates catabolic pathways and, at the same time, inhibits the rate of anabolic reactions to regain the correct energy charge for adenylates.

Capsaicin treatment of human colorectal adenocarcinoma (HT29) cell line is shown to cause AMPK activation and inhibition of acetyl-CoA carboxylase (ACC), a well-known AMPK substrate, which indicates capsaicin inhibits lipid biosynthesis. The above concepts were implicated in apoptosis caused by capsaicin.

3.3 Capsaicin and cell cycle

The cycle of cells is a series of stages that cells undergo to allow them to divide and produce new cells. The cell cycle is divided into phases G0/G1, S, and G2/M stages. The cyclins, cyclin-dependent kinases (CDKs), and the CDK inhibitors are essential parts of the cell cycle. There are DNA checkpoints to assure DNA replication integrity. Such checkpoints and repair pathways render cellular responses to damage to DNA easier [14]. When stimulated, the CDKs provide the cells with a driving force to pass from one stage to the next, but if cyclin and/or CDKs are impaired, cell cycle arrest [17, 33, 34] happens. Thus, any alteration in these pathways raises the cancer risk. It was reported that capsaicin inhibits CDK2, CDK4, and CDK6 inhibiting the proliferation of 5637 bladder carcinoma cells via cycle arrest [34].

3.4 Capsaicin and p53

The p53 tumor suppressor prevents cell proliferation by inducing cell cycle arrest and apoptosis in response to cell stress such as damage to DNA, hypoxia, and activation of oncogenes. Phosphorylation is critical for p53-dependent transactivation at the Ser-15 residue [17]. P53 promotes apoptosis by a linear pathway involving Bax transactivation, cytosol-to-membrane Bax translocation, mitochondrial cytochrome c release, and caspase-9 activation, followed by caspase-3, -6, and -7 activation. Research studies indicate p53 is a target of capsaicin's anticancer action. Capsaicin was found to induce p53 phosphorylation of residue Ser-15 and enhanced p53 acetylation by sirtuin 1 downregulation, which is responsible for the initiation of apoptosis [35].

Incubated adenocarcinoma gastric cell line (AGS) of human gastric cancer with different capsaicin concentrations in the presence and absence of p53 siRNA showed that capsaicin induces apoptosis via p53 upregulation in AGS cells and that the apoptotic effect of capsaicin is p53-dependent [36]. It also observed that the tendency of capsaicin to induce the expression of proapoptotic proteins, such as Bax, caspase-3, and caspase-8, was almost entirely diminished by hitting down p53. Effects of capsaicin on the same cell type reported that caspase-3 activity increased with capsaicin exposure, indicating that capsaicin may serve as an anti-tumorigenic agent in human gastric cancer [37].

3.5 Capsaicin and β-catenin

β-catenin is a beneficial 90 kD protein that leads to cell growth under normal physiological conditions. It is a key transcription factor in the signaling of Wingless-Int (Wnt) and plays a vital role in stem cell regeneration and organ regrowth. Abnormal expression of β-catenin causes the malignant conversion of normal cells, and its anomalous activity has been documented in several cancer types. β-catenin is essentially active in many types of cancer cells. In a recent study, it was reported that capsaicin downregulated β-catenin transcription, decreased its protein stability, and caused apoptosis of colorectal cancer cells [38].

3.6 Capsaicin and angiogenesis

Angiogenesis is the development of new blood vessels from preexisting vasculature as well as an important homeostatic process for normal wound healing and embryonic growth. It involves the stimulation of endothelial cells, cell proliferation, invasion, chemotactic migration, and differentiation into new blood vessels [39]. Angiogenesis is a central player in cancer cell growth and tumor metastasis. The development of an angiogenic phenotype is regarded as a vital step in the progression of tumors [40, 41]. The cancer cells grew to 1–2 mm^3 in thickness in absence of circulation and stopped, but in the angiogenic area, they expanded beyond 2 mm^3. Tumors can become necrotic or even apoptotic in the absence of vascular support [42]. Thus, angiogenesis is a significant factor in cancer progression.

Some factors such as growth factors, cytokines, and vascular endothelial growth factor (VGEF) regulate angiogenesis. Nevertheless, VEGF plays a significant role in angiogenesis. In addition, endothelial cells produce growth factors that induce autocrine and paracrine growth of tumors. The initiation of angiogenesis correlates with the increased entry into the circulation of neoplastic cells and thus promotes metastasis [43]. Treatment of endothelial cells with capsaicin blocked the sprouting and development of VEGF-induced vessels in Matrigel mouse assay that was correlated with downregulation of p38 mitogen activated protein kinases (MAPK), protein kinase B (PKB), and focal adhesion kinase (FAK) activation [44, 45].

Furthermore, capsaicin enhanced deterioration of hypoxia-inducible factor1α, which is a crucial transcription factor rising VEGF transcription. Capsaicin thus interferes with typical angiogenic signaling pathways and can have the ability to suppress cancer becoming a malignant one. It was reported that capsaicin's anti-angiogenic activity was associated with reduced cyclin D1 expression, which results in decreased Rb phosphorylation, leading to the GI arrest of human endothelial cells of the umbilical cord (HUVEC) [46]. In addition, capsaicin also inhibited focal adhesion kinase (FAK) and p38 kinase activation caused by VEGF. All of the above results indicate that capsaicin is a dietary anti-angiogenic agent.

3.7 Capsaicin and metastasis

Cancer metastasis is the hallmark of tumor malignancy, which starts with the spread of cancer cells from the principal tumor to nearby tissues and distant organs, and it is the main cause of cancer morbidity and mortality. Metastatic cancer is resistant to treatment and contributes 80% of cancer-related deaths, which remains a great problem for cancer therapy [47]. Invasion and movement of tumor cells include the proteolytic degradation of extracellular matrix components by tumor cell-secreted proteases, involving serine proteases, plasminogen activators, and matrix metalloproteinases (MMPs).

Cancer type	Cell line	Effective doses (μM)	Anticancer mechanism
Human colorectal cancer	HCT 116	100–200	Induced autophagy
	LoVo, SW480	100	Induced anti-tumorigenesis; deregulation of β-catenin/TCF-dependent signaling
	Colo 205	150	Induced cell death, increased ROS, and proapoptotic proteins
Human breast cancer	MCF-7	50–300	Induced autophagy, inhibited growth, and induced apoptosis
	T47D, BT-474, SKBR-3	200	Inhibited growth and increased apoptosis
	MDA-MB231	20–200	Induced apoptosis and dysfunctions in mitochondria. Antiproliferative activity and arrest of cell cycle into G2/M phase. Enhances the apoptotic effects of TRIAL by activating the calcium-CaMKII-Sp1 pathway
Human prostate cancer	LNCaP	40–50	Inhibited proliferation and induced apoptosis
	PC-3	20–50	
	DU-145	500	
	RWPE-1	40	
Human myeloid leukemia	HL-60	>50	Induced G0/G1 phase cell cycle arrest and apoptosis
	U937, THP-1	200	Enhances the apoptotic effects of TRIAL by activating the calcium-CaMKII-Sp1 pathway
Human esophageal epidermoid carcinoma	CE 81 T/VGH	100	Induced apoptosis and G0/G1 phase cell cycle arrest
Human melanoma	A375	100	Inhibited cell growth and promoted apoptosis
Human KB cancer cells	KB cells	150–200	Reduced cell proliferation and viability. Induced cell death and cell cycle arrest in G2/M phase
Mouse melanoma	B16-F10	50	Inhibited cell migration. Induced apoptosis
Human pancreatic cancer	AsPC-1, BxPC-3	150	Inhibited proliferation. Induced apoptosis and generated ROS
	PANC-1	200	Induced G0/G1 phase cell cycle arrest and apoptosis; and inhibited growth
Human multiple myeloma	U266, MM.1S	>5	Inhibited cell proliferation, caused accumulation of cells in G1 phase
Human hepatoma	Hep G2	10–200	Decreased cell viability, generated ROS and activated caspase-3; and induced apoptosis and autophagy
	Hep3B	200	Enhances the apoptotic effects of TRIAL by activating the calcium-CaMKII-Sp1 pathway
Human nasopharyngeal carcinoma	NPC-TW 039	200–400	Induced G0/G1 phase arrest and apoptosis. Increased ROS
Human gastric carcinoma	SMC-1	200	Induced apoptosis
Human bladder cancer	T24	100	Induced ROS production and mitochondrial membrane depolarization

Cancer type	Cell line	Effective doses (μM)	Anticancer mechanism
Human small cell lung cancer	NCI-H69, NCI-H82, DMS53, DMS114	50	Suppressed growth in all four cell lines

TCF: T-cell factor; ROS: reactive oxygen species; TRIAL: TNF-related apoptosis-inducing ligand; CaMKII-Sp1: calcium-calmodulin-dependent kinase II signaling pathway I.

Table 1.
Anticancer potential of capsaicin against various cancer cell lines.

Capsaicin revealed its anti-invasive and anti-migratory activity by modulating signaling pathways including cell invasion and migration and suppressing advanced cancer stages. Treatment with capsaicin significantly reduced the metastatic burden in transgenic adenocarcinoma of the mouse prostate (TRAMP) mice. It has been reported that it significantly inhibited the migration of melanoma cells without leading to apparent cellular cytotoxicity [48]. This effect was associated with the downregulation of the signaling cascade of phosphoinositide 3-kinase (PI3 K) and reduction of the RAS-related c3 botulinum toxin substrate 1 (RAC1), which in itself is the main kinase controlling motility and migration of cells. Capsaicin blocked the invasion and migration of human fibrosarcoma cells triggered by epidermal growth factor (EGF) by downregulation of AKT/FAK, extracellular signal-regulated kinases, and p38 MAPK signaling, and subsequent downregulation of matrix metallopeptidase 9 (MMP9) in invasive fibrosarcoma cells.

Thereby, capsaicin recruits several mechanisms of signaling for controlling migration and invasion. These include epithelial-mesenchymal transition (EMT) activation, AMP-activated protein kinase (AMPK), MMP signaling pathway, intracellular calcium elevation, VEGF, Wnt-Hedgehog control, tumor-associated NADH oxidase (tNOX), protein kinase B (Akt), MMPs, epidermal growth factor receptor (EGFR), extracellular signal-regulated kinase (ERK), p38 MAP kinase, RAC1, nuclear factor "kappa-light-chain-enhancer" of activated B-cells (NF-kB), and AP-1 [49–54]. **Table 1** depicts the anticancer potential of capsaicin against various cancer cell lines.

4. Synergistic anticancer activity of capsaicin with other compounds

Recent development has allowed synergistic drugs to treat a wide array of cancers. Combinations of low doses of chemopreventive agents have progressively been used clinically in the treatment of various cancers in the present days. Compared to the traditional single drug policy, the combination strategy increases therapeutic effects or reduces drug resistance [55]. Computational progress in qualitatively testing and predicting synergistic components has been experienced in the last few years [56–58]. Multiple studies have illustrated that novel combination therapies with different phytochemicals and chemopreventive drugs can elicit increased antitumor activity through additive or synergistic action [59, 60].

The pathways can include parallel action on cancer suppression by modifying carcinogen detoxification and metabolism of hormones, scavenging oxidative stress, and enhancing immunity [61–63]. Hundreds of phytochemicals have various physicochemical and kinetic properties that target different signaling pathways and transcription factors that decide the phenotype of cancer. The use of several bioac-tive compounds with distinct anticancer pathways could, therefore, be a promising method for successful cancer treatment and prevention.

Capsaicin and other compounds are known to have synergistic anticancer properties. Combined with resveratrol, capsaicin facilitated apoptosis by elevating nitric oxide (NO) via a p53-dependent manner [64]. It has synergistic anticancer activity with pirarubicin, an anthracycline drug, by activating TRPV1 in bladder cancer [65]. Capsaicin shows a synergistic effect with the dietary phytoestrogen, genistein by regulation of AMPK and cyclooxygenase 2 in breast cancer cells [66]. Capsaicin and 3,3′-diindolylmethane, a key in vivo metabolite of indole-3-carbinol, abundantly present in cruciferous vegetables, have recently been reported to work synergistically to induce apoptosis in colorectal cancer, by altering the transcriptional function of the nuclear factor kappa B, p53, and control apoptosis-related genes [67]. Capsaicin, and brassinin, a form of indole derived from cruciferous vegetables, showed synergistic anticancer activity by suppressing MMP2 and -9 expression and enzymatic activities, and invasion, and migration of prostate carcinoma cells [68]. Sometimes, capsaicin also interacts with chemotherapy drugs. For example, in some myeloid leukemia cells [69], capsaicin enhanced the therapeutic efficacy with 12-O-Tetradecanoylphorbol-13-acetate. Capsaicin can, surprisingly, affect the viability of cancer stem cells. Capsaicin has been found to cause cancer stem cell death by inhibiting the NOTCH signaling pathway in stem cells of breast cancer [70].

5. Conclusion

An emerging cancer research area is the search for suitable molecular targets and effective anticancer compounds that modify the cancer targets. Capsaicin exhibits significant anticancer activity in various tumor stages by targeting multiple signaling pathways and cancer-associated genes. As a whole, capsaicin's anticancer pathways involve apoptosis initiation, cell growth arrest, and suppression of angio-genesis and metastasis. Capsaicin and its analogs activate the tumor-suppressive signaling pathway and associated transcription factors while inhibiting oncogenic signaling pathways and cancer cell promoters. In addition, capsaicin acts synergisti-cally with other anticancer agents, allowing for the potential use of capsaicin with other chemotherapeutic agents in cancer therapy, and shows double advantage, i.e., capsaicinoids enhances the chemotherapeutic effect and pain relief for cancer patients. More research on capsaicin's and its analogs as anticancer targets holds the potential for future treatments and requires more study to improve our understanding of its efficacy in cancer treatment and prevention.

Author details

Balasubramanian Arul* and Ramalingam Kothai
Vinayaka Mission's College of Pharmacy, Vinayaka Mission's Research Foundation (Deemed to be University), Salem, Tamil Nadu, India

*Address all correspondence to: arul1971@yahoo.com

References

[1] Dong X, Li X, Ding L, Cui F, Tang Z, Liu Z. Stage extraction of capsaicinoids and red pigments from fresh red pepper (*Capsicum*) fruits with ethanol as solvent. LWT - Food Science and Technology [Internet]. 2014;**59**(1):396-402. DOI: 10.1016/j.lwt.2014.04.051

[2] Govindarajan VS, Salzer UJ. *Capsicum*-production, technology, chemistry, and quality part 1: History, botany, cultivation, and primary processing. Critical Reviews in Food Science and Nutrition [Internet]. 1985;**22**(2):109-176. DOI: 10.1080/10408398509527412

[3] Szolcsanyi J. Capsaicin and sensory neurones: A historical perspective. Progress in Drug Research. 2014;**68**:1-37. DOI: 10.1007/978-3-0348-0828-6_1

[4] Kehie M, Kumaria S, Tandon P. Manipulation of culture strategies to enhance capsaicin biosynthesis in suspension and immobilized cell cultures of *Capsicum chinense* Jacq. cv. Naga King Chili. Bioprocess and Biosystems Engineering. 2014;**37**(6):1055-1063. DOI: 10.1007/s00449-013-1076-2

[5] Bernal MA, Calderon AA, Pedreno MA, Ferrer MA, Merino de Caceres F, Ros Barcelo A. Purification of a basic peroxidase isoenzyme from *Capsicum* fruits and the immunoinhibition of its capsaicin oxidation capacity by antibodies raised against horseradish peroxidase. Zeitschrift für Lebensmittel-Untersuchung und - Forschung. 1994;**199**(3):240-242. DOI: 10.1007/bf01193453

[6] Walpole CS, Bevan S, Bloomfield G, Breckenridge R, James IF, Ritchie T, et al. Similarities and differences in the structure-activity relationships of capsaicin and resiniferatoxin analogues. Journal of Medicinal Chemistry. 1996;**39**(15):2939-2952. DOI: 10.1021/jm960139d

[7] O'Neill J, Brock C, Olesen AE, Andresen T, Nilsson M, Dickenson AH. Unravelling the mystery of capsaicin: A tool to understand and treat pain. Pharmacological Reviews. 2012;**64**(4):939-971. DOI: 10.1124/pr.112.006163

[8] Castillo E, Lopez-Gonzalez I, De Regil-Hernandez R, Reyes-Duarte D, Sanchez-Herrera D, Lopez-Munguia A, et al. Enzymatic synthesis of capsaicin analogs and their effect on the T-type Ca2+ channels. Biochemical and Biophysical Research Communications. 2007;**356**(2):424-430. DOI: 10.1016/j.bbrc.2007.02.144

[9] Maoka T, Mochida K, Kozuka M, Ito Y, Fujiwara Y, Hashimoto K, et al. Cancer chemopreventive activity of carotenoids in the fruits of red paprika *Capsicum annuum* L. Cancer Letters. 2001;**172**(2):103-109. DOI: 10.1016/s0304-3835(01)00635-8

[10] Matsumoto J, Takeuchi K, Okabe S. Characterization of gastric mucosal blood flow response induced by intragastric capsaicin in rats. Japanese Journal of Pharmacology. 1991;**57**(2):205-213. DOI: 10.1254/jjp.57.205

[11] Wallace M, Pappagallo M. Qutenza(R): A capsaicin 8% patch for the management of postherpetic neuralgia. Expert Review of Neurotherapeutics. 2011;**11**(1):15-27. DOI: 10.1586/ern.10.182

[12] Salat K, Jakubowska A, Kulig K. Zucapsaicin for the treatment of neuropathic pain. Expert Opinion on Investigational Drugs. 2014;**23**(10):1433-1440. DOI: 10.1517/13543784.2014.956079

[13] Chapa-Oliver AM, Mejía- Teniente L. Capsaicin: From plants to a cancer-suppressing agent. Molecules [Internet]. 2016;**21**(8):931. DOI: 10.3390/molecules21080931

[14] Aggarwal BB, Shishodia S. Molecular targets of dietary agents for prevention and therapy of cancer. Biochemical Pharmacology. 2006;**71**(10):1397-1421. DOI: 10.1016/j.bcp.2006.02.009

[15] Hanahan D, Weinberg RA. Hallmarks of cancer: The next generation. Cell. 2011;**144**(5):646-674. DOI: 10.1016/j.cell.2011.02.013

[16] Clark R, Lee S-H. Anticancer properties of capsaicin against human Cancer. Anticancer Research. 2016;**36**(3):837-843

[17] Oyagbemi AA, Saba AB, Azeez OI. Capsaicin: A novel chemopreventive molecule and its underlying molecular mechanisms of action. Indian Journal of Cancer. 2010;**47**(1):53-58. DOI: 10.4103/0019-509X.58860

[18] Bley K, Boorman G, Mohammad B, McKenzie D, Babbar S. A comprehensive review of the carcinogenic and anticarcinogenic potential of capsaicin. Toxicologic Pathology. 2012;**40**:847-873. DOI: 10.1177/0192623312444471

[19] Chalah A, Khosravi-Far R. The mitochondrial death pathway. Advances in Experimental Medicine and Biology. 2008;**615**:25-45. DOI: 10.1007/978-1-4020-6554-5_3

[20] Amantini C, Ballarini P, Caprodossi S, Nabissi M, Morelli MB, Lucciarini R, et al. Triggering of transient receptor potential vanilloid type 1 (TRPV1) by capsaicin induces Fas/CD95-mediated apoptosis of urothelial cancer cells in an ATM-dependent manner. Carcinogenesis. 2009;**30**(8):1320-1329. DOI: 10.1093/carcin/bgp138

[21] Kim SR, Kim SU, Oh U, Jin BK. Transient receptor potential vanilloid subtype 1 mediates microglial cell death In vivo and In vitro via Ca2+ mediated mitochondrial damage and cytochrome C release. Journal of Immunology [Internet]. 2006;**177**(7):4322-4329. DOI: 10.4049/jimmunol.177.7.4322

[22] Caprodossi S, Amantini C, Nabissi M, Morelli MB, Farfariello V, Santoni M, et al. Capsaicin promotes a more aggressive gene expression phenotype and invasiveness in null-TRPV1 urothelial cancer cells. Carcinogenesis. 2011;**32**(5):686-694. DOI: 10.1093/carcin/bgr025

[23] Amantini C, Mosca M, Nabissi M, Lucciarini R, Caprodossi S, Arcella A, et al. Capsaicin-induced apoptosis of glioma cells is mediated by TRPV1 vanilloid receptor and requires p38 MAPK activation. Journal of Neurochemistry. 2007;**102**(3):977-990. DOI: 10.1111/j.1471-4159.2007.04582.x

[24] Lau JK, Brown KC, Dom AM, Witte TR, Thornhill BA, Crabtree CM, et al. Capsaicin induces apoptosis in human small cell lung cancer via the TRPV6 receptor and the calpain pathway. Apoptosis. 2014;**19**(8):1190-1201. DOI: 10.1007/s10495-014-1007-y

[25] Zhang R, Humphreys I, Sahu RP, Shi Y, Srivastava SK. In vitro and in vivo induction of apoptosis by capsaicin in pancreatic cancer cells is mediated through ROS generation and mitochondrial death pathway. Apoptosis. 2008;**13**(12):1465-1478. DOI: 10.1007/s10495-008-0278-6

[26] Hail NJ, Lotan R. Examining the role of mitochondrial respiration in vanilloid-induced apoptosis. Journal of the National Cancer Institute. 2002;**94**(17):1281-1292. DOI: 10.1093/jnci/94.17.1281

[27] Diaz-Laviada I. Effect of capsaicin on prostate cancer cells. Future

Oncology. 2010;**6**(10):1545-1550. DOI: 10.2217/fon.10.117

[28] Datta P, Pramanik KC, Mehrotra S, Srivastava SK. Capsaicin media ted oxidative stress in pancreatic cancer. In: Cancer. London, UK: Elsevier; 2014. pp. 241-246

[29] Sinha K, Das J, Pal PB, Sil PC. Oxidative stress: The mitochondria-dependent and mitochondria-independent pathways of apoptosis. Archives of Toxicology. 2013;**87**(7):1157-1180. DOI: 10.1007/s00204-013-1034-4

[30] Surh Y-J. More than spice: Capsaicin in hot chili peppers makes tumor cells commit suicide. JNCI: Journal of the National Cancer Institute [Internet]. 2002;**94**(17):1263-1265. DOI: 10.1093/jnci/94.17.1263

[31] Kuhajda FP. AMP-activated protein kinase and human cancer: Cancer metabolism revisited. International Journal of Obesity. 2008;**32**(Suppl 4):S36-S41. DOI: 10.1038/ijo.2008.121

[32] Kim YM, Hwang J-T, Kwak DW, Lee YK, Park OJ. Involvement of AMPK signaling cascade in capsaicin-induced apoptosis of HT-29 colon cancer cells. Annals of the New York Academy of Sciences. 2007;**1095**:496-503. DOI: 10.1196/annals.1397.053

[33] Hardie DG, Alessi DR. LKB1 and AMPK and the cancer-metabolism link —Ten years after. BMC Biology. 2013;**11**:36. DOI: 10.1186/1741-7007-11-36

[34] Chen D, Yang Z, Wang Y, Zhu G, Wang X. Capsaicin induces cycle arrest by inhibiting cyclin-dependent-kinase in bladder carcinoma cells. International Journal of Urology. 2012;**19**(7):662-668. DOI: 10.1111/j.1442-2042.2012.02981.x

[35] Ito K, Nakazato T, Yamato K, Miyakawa Y, Yamada T, Hozumi N, et al. Induction of apoptosis in leukemic cells by homovanillic acid derivative, capsaicin, through oxidative stress: Implication of phosphorylation of p53 at Ser-15 residue by reactive oxygen species. Cancer Research. 2004;**64**(3):1071-1078. DOI: 10.1158/0008-5472.can-03-1670

[36] Sarkar A, Bhattacharjee S, Mandal DP. Induction of apoptosis by eugenol and capsaicin in human gastric cancer AGS cells—Elucidating the role of p53. Asian Pacific Journal of Cancer Prevention. 2015;**16**(15):6753-6759. DOI: 10.7314/apjcp.2015.16.15.6753

[37] Park S-Y, Kim J-Y, Lee S-M, Jun C-H, Cho S-B, Park C-H, et al. Capsaicin induces apoptosis and modulates MAPK signaling in human gastric cancer cells. Molecular Medicine Reports. 2014;**9**(2):499-502. DOI: 10.3892/mmr.2013.1849

[38] Lee S-H, Richardson RL, Dashwood RH, Baek SJ. Capsaicin represses transcriptional activity of beta-catenin in human colorectal cancer cells. The Journal of Nutritional Biochemistry. 2012;**23**(6):646-655. DOI: 10.1016/j.jnutbio.2011.03.009

[39] Potente M, Gerhardt H, Carmeliet P. Basic and therapeutic aspects of angiogenesis. Cell. 2011;**146**(6):873-887. DOI: 10.1016/j.cell.2011.08.039

[40] Folkman J. Role of angiogenesis in tumor growth and metastasis. Seminars in Oncology. 2002;**29**(6 Suppl 16):15-18. DOI: 10.1053/sonc.2002.37263

[41] Lin Z, Zhang Q , Luo W. Angiogenesis inhibitors as therapeutic agents in cancer: Challenges and future directions. European Journal of Pharmacology. 2016;**793**:76-81. DOI: 10.1016/j.ejphar.2016.10.039

[42] Holmgren L, O'Reilly MS, Folkman J. Dormancy of micrometastases: Balanced proliferation and apoptosis in the presence of angiogenesis suppression. Nature

Medicine. 1995;**1**(2):149-153. DOI: 10.1038/nm0295-149

[43] Parangi S, O'Reilly M, Christofori G, Holmgren L, Grosfeld J, Folkman J, et al. Antiangiogenic therapy of transgenic mice impairs de novo tumor growth. Proceedings of the National Academy of Sciences of the United States of America [Internet]. 1996;**93**(5):2002-2007. DOI: 10.1073/pnas.93.5.2002

[44] Bhutani M, Pathak AK, Nair AS, Kunnumakkara AB, Guha S, Sethi G, et al. Capsaicin is a novel blocker of constitutive and interleukin-6-inducible STAT3 activation. Clinical Cancer Research. 2007;**13**(10):3024-3032. DOI: 10.1158/1078-0432.CCR-06-2575

[45] Chakraborty S, Adhikary A, Mazumdar M, Mukherjee S, Bhattacharjee P, Guha D, et al. Capsaicin-induced activation of p53-SMAR1 auto-regulatory loop down-regulates VEGF in non-small cell lung cancer to restrain angiogenesis. PLOS One. 2014;**9**(6):e99743. DOI: 10.1371/journal.pone.0099743

[46] Min J-K, Han K-Y, Kim E-C, Kim Y-M, Lee S-W, Kim O-H, et al. Capsaicin inhibits in vitro and in vivo angiogenesis. Cancer Research. 2004;**64**(2):644-651. DOI: 10.1158/0008-5472.can-03-3250

[47] Talmadge JE, Fidler IJ. AACR centennial series: The biology of cancer metastasis - Historical perspective. Cancer Research. 2010;**70**(14):5649-5669. DOI: 10.1158/0008-5472.CAN-10-1040

[48] Shin DH, Kim OH, Jun HS, Kang MK. Inhibitory effect of capsaicin on B16-F10 melanoma cell migration via the phosphatidylinositol 3-kinase/Akt/Rac1 signal pathway. Experimental & Molecular Medicine. 2008;**40**(5):486-494. DOI: 10.3858/emm.2008.40.5.486

[49] Hurley JD, Akers AT, Friedman JR, Nolan NA, Brown KC, Dasgupta P. Non-pungent long chain capsaicin-analogs arvanil and olvanil display better anti-invasive activity than capsaicin in human small cell lung cancers. Cell Adhesion & Migration [Internet]. 2017;**11**(1):80-97. DOI: 10.1080/19336918.2016.1187368

[50] Amantini C, Morelli MB, Nabissi M, Cardinali C, Santoni M, Gismondi A, et al. Capsaicin triggers autophagic cell survival which drives epithelial mesenchymal transition and chemoresistance in bladder cancer cells in an Hedgehog-dependent manner. Oncotarget. 2016;**7**(31):50180-50194. DOI: 10.18632/oncotarget.10326

[51] Lee G-R, Jang SH, Kim CJ, Kim A-R, Yoon D-J, Park N-H, et al. Capsaicin suppresses the migration of cholangiocarcinoma cells by down-regulating matrix metalloproteinase-9 expression via the AMPK-NF-kappaB signaling pathway. Clinical & Experimental Metastasis. 2014;**31**(8):897-907. DOI: 10.1007/s10585-014-9678-x

[52] Lin M-H, Lee Y-H, Cheng H-L, Chen H-Y, Jhuang F-H, Chueh PJ. Capsaicin inhibits multiple bladder Cancer cell phenotypes by inhibiting tumor-associated NADH oxidase (tNOX) and sirtuin1 (SIRT1). Molecules. 2016;**21**(7). DOI: 10.3390/molecules21070849

[53] Xu S, Zhang L, Cheng X, Yu H, Bao J, Lu R. Capsaicin inhibits the metastasis of human papillary thyroid carcinoma BCPAP cells through the modulation of the TRPV1 channel. Food & Function. 2018;**9**(1):344-354. DOI: 10.1039/c7fo01295k

[54] Wutka A, Palagani V, Barat S, Chen X, El Khatib M, Gotze J, et al. Correction: Capsaicin treatment attenuates cholangiocarcinoma carcinogenesis. PlOS One, United States. 2016;**11**:e0162673. DOI: 10.1371/journal. pone.0162673

[55] Chou T-C. Theoretical basis, experimental design, and computerized simulation of synergism and antagonism in drug combination studies. Pharmacological Reviews. 2006;**58**(3):621-681. DOI: 10.1124/pr.58.3.10

[56] Han K, Jeng EE, Hess GT, Morgens DW, Li A, Bassik MC. Synergistic drug combinations for cancer identified in a CRISPR screen for pairwise genetic interactions. Nature Biotechnology. 2017;**35**(5):463-474. DOI: 10.1038/nbt.3834

[57] Sarah C. Identifying synergistic drug combinations. Nature Reviews. Drug Discovery [Internet]. 2017;**16**(5):314. DOI: 10.1038/nrd.2017.76

[58] Sheng Z, Sun Y, Yin Z, Tang K, Cao Z. Advances in computational approaches in identifying synergistic drug combinations. Briefings in Bioinformatics. 2018;**19**(6):1172-1182. DOI: 10.1093/bib/bbx047

[59] Sarkar FH, Li Y. Using chemopreventive agents to enhance the efficacy of cancer therapy. Cancer Research. 2006;**66**(7):3347-3350. DOI: 10.1158/0008-5472.CAN-05-4526

[60] Liu RH. Potential synergy of phytochemicals in cancer prevention: Mechanism of action. The Journal of Nutrition. 2004;**134**(12 Suppl):3479S-3485S. DOI: 10.1093/jn/134.12.3479S

[61] Dragsted LO, Strube M, Larsen JC. Cancer-protective factors in fruits and vegetables: Biochemical and biological background. Pharmacology & Toxicology. 1993;**72**(Suppl 1):116- 135. DOI: 10.1111/j.1600-0773.1993.tb01679.x

[62] Liu RH. Health benefits of fruit and vegetables are from additive and synergistic combinations of phytochemicals. The American Journal of Clinical Nutrition. 2003;**78**(3 Suppl): 517S-520S. DOI: 10.1093/ajcn/78.3.517S

[63] de Kok TM, van Breda SG, Manson MM. Mechanisms of combined action of different chemopreventive dietary compounds: A review. European Journal of Nutrition. 2008;**47**(Suppl 2):51-59. DOI: 10.1007/s00394-008-2006-y

[64] Kim MY, Trudel LJ, Wogan GN. Apoptosis induced by capsaicin and resveratrol in colon carcinoma cells requires nitric oxide production and caspase activation. Anticancer Research. 2009;**29**(10):3733-3740

[65] Zheng L, Chen J, Ma Z, Liu W, Yang F, Yang Z, et al. Capsaicin enhances anti-proliferation efficacy of pirarubicin via activating TRPV1 and inhibiting PCNA nuclear translocation in 5637 cells. Molecular Medicine Reports. 2016;**13**(1):881-887

[66] Hwang J-T, Lee Y-K, Shin J-I, Park OJ. Anti-inflammatory and anticarcinogenic effect of genistein alone or in combination with capsaicin in TPA-treated rat mammary glands or mammary cancer cell line. Annals of the New York Academy of Sciences. 2009;**1171**:415-420. DOI: 10.1111/j.1749-6632.2009.04696.x

[67] Clark R, Lee J, Lee S-H. Synergistic anticancer activity of capsaicin and 3,3′-diindolylmethane in human colorectal cancer. Journal of Agricultural and Food Chemistry. 2015;**63**(17):4297-4304. DOI: 10.1021/jf506098s

[68] Kim S-M, Oh EY, Lee JH, Nam D, Lee SG, Lee J, et al. Brassinin combined with capsaicin enhances apoptotic and anti-metastatic effects in PC-3 human prostate Cancer cells. Phytotherapy Research. 2015;**29**(11):1828-1836. DOI: 10.1002/ptr.5478

[69] Zheng X, Ryan A, Patel N, Klemons S, Hansson A, Shih WJ, et al.

Synergistic stimulatory effect of 12-O-tetradecanoylphorbol-13-acetate and capsaicin on macrophage differentiation in HL-60 and HL-525 human myeloid leukemia cells. International Journal of Oncology. 2005;**26**(2):441-448

[70] Shim Y, Song JM. Quantum dot nanoprobe-based high-content monitoring of notch pathway inhibition of breast cancer stem cell by capsaicin. Molecular and Cellular Probes. 2015;**29**(6):376-381. DOI: 10.1016/j.mcp.2015.09.004

8

The Effects of Consumption of *Capsicum* on Some Neurobehavioural Parameters

Eme Efiom Osim, Anyauba Nmaju, Azubuike Nwankwo, Udemeobong Okon and Joshua Iwasam

Abstract

Capsicum annum, an extensively cultivated vegetable, is commonly used to spice many dishes prepared in several parts of the world. It contains capsaicinoids, which give it a characteristic pungency. The most active and well known amongst these capsaicinoids is capsaicin (8-methyl-N-vanillyl-6-nonenamide), which is neurogenic and may affect neuronal functions. Therefore, our study investigated the effects of consumption of chilli pepper and capsaicin diets on neurobehaviour of CD-1 mice. The neurobehavioural parameters assessed were anxiety, motor coordination, pain, social behaviour, learning and memory. The animals were randomly assigned into three groups of 10 mice each, namely, control, pepper-diet (20% w/w), and capsaicin-diet (10% w/w) groups. Their learning and memory abilities were assessed through their ability to locate the hidden platform model of Morris water maze apparatus. The elevated plus maze and light-dark transition box were used to assess anxiety-related behaviour, while the beam walking test and nesting behaviour were used to determine motor coordination and social behaviour, respectively. Tail immersion, hot plate, and formalin tests were conducted to assess pain perception. Consumption of the chilli pepper and capsaicin diets decreased pain perception, increased anxiety, and impaired learning and memory but enhanced social behaviour and motor coordination in mice.

Keywords: *Capsicum annum*, capsaicin, pain, social behaviour, anxiety, motor coordination, learning and memory

1. Introduction

Spices are used all over the world not only for their flavours but also for their medicinal properties [1]. One of such spices is chilli pepper. Chilli pepper or chillies are vegetables used world over for culinary purposes and belong to the family Solanaceae [2].

These peppers are widely used as spices in food industry and in a broad variety of medicinal applications worldwide [3]. These spices are remarkable sources of antioxidant compounds including phenolic compounds and flavonoids, of which their consumption has potential health benefits due to their activity as free radical scavengers and may also help prevent inflammatory diseases and pathologies associated with oxidative damage such as atherosclerosis and Alzheimer's disease [3].

Chilli peppers are called *Ntokon* in Efik, *Osé* in Ibo and *Brukunu* in Hausa languages in Nigeria. They are usually red or green in colour. It is the most commonly consumed pepper in Nigeria.

The substances that give chillies their hot sensation and intensity when ingested or contacted are pungent chemical compounds collectively known as Capsaicinoids with Capsaicin (trans-8-methyl-N-vanillyl-6-nonenamide) as the most abundant [4]. Exposure to the skin causes intense burning sensation while exposure to the eyes leads to intense tearing, conjunctivitis and blepharospasm [5].

The burning sensations associated with Capsaicin ingestion result from the activation of transient receptor potential, vanilloid 1 (TRPV1) located in the gut and other organs [6]. The stimulation of these TRPV1 receptors brings about the influx of sodium and calcium ions which results in the depolarization of nociceptive neurons, leading to the firing of action potentials and finally the sensation of spiciness [7]. Capsaicin is an alkaloid which is highly volatile, odourless, colourless and hydrophobic [8]. It structurally, belongs to a group of chemicals known as vanilloids, having a vanillyl (methylcatechol) head group (A-region) and an aliphatic tail (hydrophobic —C-region) linked by a central amide bond (B-region). This combination of these regions is accountable for the pharmacological activities of capsaicin [8].

Numerous health benefits are believed to emanate from Chilli pepper consumption [7]. Kempaiah et al. [9] reported that Capsaicin demonstrated protective effects against obesity and cholesterol by speeding up metabolism through stress hormone release. It is used as a topical agent in formulations against arthritis [10] and also in preparation of defensive sprays because of its irritant properties and ability to cause neurogenic inflammation (stinging sensation of hands, eyes and mouth) [11, 12].

Capsicum annum contains capsaicin which is neurogenic [11] and therefore, can affect neuronal activities in the body [13]. It is therefore conceivable that it may affect some nervous function parameters such as pain, social behaviour, anxiety, motor coordination, learning and memory. Hence, we present the report of our study that investigated the comparative effects of long-term consumption of capsaicin and chillies on the aforementioned neurobehavioural parameters using mice as experimental models to ascertain whether the effects obtained with pepper-diet consumption can be attributed to capsaicin.

2. Materials and methods

2.1 Preparation and storage of experimental diets

Half-washed basin of fresh red chilli pepper (*Capsicum annum*) was procured from a local market in Nigeria. It was washed and sun-dried for 4 days. The dried samples were then pulverised using electric blender to obtain a fine powder. The pepper powder was then stored in air-tight rubber container from which pepper diets were prepared.

Capsaicin (95% pure) was obtained from Wuxi Gorunjie natural-Pharma Co. Ltd., Jiangsu China.

2.2 Animal treatment

Thirty (30) male mice of CD-1 strain weighing between 22 and 34 g were used for the study. They were kept in a well-ventilated room under room temperature (25 ± 2°C), humidity of 8 ± 5% and 12/12 h light/dark cycle and allowed 1 week for acclimatisation to the research environment before the experiments. They were randomly assigned into three groups, namely; control group that received normal rodent chow, pepper group that were fed 20% chilli pepper diet and capsaicin group that were given 10% capsaicin diet. Each group comprised 10 mice. Each mouse was allowed drinking water *ad libitum*. This treatment was done for 28 days and within this period, their beddings, feed and water were hygienically handled and changed daily.

2.3 Approval for animal use

Approval for this study was obtained from the Faculty Animal Ethics Committee of Faculty of Basic Medical Sciences, University of Calabar with Protocol number 014PY20314.

2.4 Behavioural protocol

2.4.1 Assessment of anxiety levels

2.4.1.1 The elevated plus maze

The elevated plus maze apparatus designed according to the description of Lister [14], and the test protocol adapted by Okon et al. [15] were used in this present study. This apparatus is used to assess the anxiety and fear levels of the mice. The test is based on the inborn aversion of rodents to open or bright illuminated spaces. The maze has two open arms ($45 \times 5\ cm^2$) with 0.25 cm high edges and two closed arms ($40 \times 5\ cm^2$) with 15 cm high walls radiating from a central square (5×5 cm). The open arms contain a slight edge (4 mm high) to prevent the mice from slipping and falling off the edge [16].

Prior to the test, the plus maze arms, surfaces and closed sides were cleaned with methylated spirit to eliminate olfactory clues and to remove faecal boll and urine. The mice were placed in the central square of the plus maze such that the mice faced an open arm away from the experimenter upon placement. Immediately after placement, a stop watch was started and the mice were allowed to explore the apparatus for 5 min. The test sessions were recorded and videotaped. Behaviours scored included open arm entry, open arm entry duration; head dip, rearing and stretch attend posture frequencies.

2.4.2 Assessment of motor coordination

2.4.2.1 Beam walking

The beam walking assesses fine motor coordination and balance [17]. This test examines the ability of the mice to remain upright and to walk on an elevated and relatively narrow beam [17]. The beam has a length of 120 cm, a width of 0.6 cm and is suspended about 60 above some foam pads. The beam is marked at 5 cm and 1 cm intervals. It is composed of wood and is coated with black paint. The mouse was placed on one end of the beam. The trial was started after the mouse has secured its grip on the beam and lasted for approximately 5 min. The tests were videotaped for scoring. The parameters scored included the number of foot slips and falls.

2.4.3 Assessment of pain perception

2.4.3.1 Tail immersion test

This test used by Ramabadran et al. [18] assesses the basic pain response in mice to thermal stimuli. This test measures spinally driven aspects of pain. Here, the animals were immobilised using tube restrainer (which also allowed free movement of the tail). The distal half of each mouse tail was immersed in water bath (at 50°C) contained in a beaker with 20 s time-out. The duration of time taken for the mice to flick its tail away from the heat of the hot water was measured in seconds. This is known as latency of flick [19].

2.4.3.2 Hot plate test

This is a test of thermal nociception, model of short duration stimuli [20]. Each mouse was exposed to a hot surface within a confined glass cage (whose temperature was maintained at 55 ± 0.5°C) for maximum duration of 30 s [21]. The time it took for each mouse to start licking its foot pad was recorded. Higher frequency and duration of paw lick indicate higher pain perception. The time taken for it to jump (latency of jump) was also recorded. The longer the latency of jump, the less pain it felt. These behaviours are the most common measures of pain threshold and are considered supra-spinally integrated [20].

2.4.4 Assessment of social behaviour

Nesting behaviour test used by Bender et al. [22] and Deacon [23] as an assay for social behaviour was used in this present study. The test was conducted in individual home cages. In total, 1 h before giving the mice nestling materials, all enrichment objects were removed from the home cages of the mice. About 3.0 g of nestling material was supplied to each mouse in its home cage and allowed for 24 h after which the nests were assessed using the rating scale supplied by Deacon [23] (**Table 1**). This assessment was based on what was seen in the home cages of the mice. Extreme care was taken while observation was carried out, as causing panic to a mouse could result in the destruction of the nest that was built.

Rating	Requirements
1	Nestlet not noticeably touched (90% or more intact)
2	Nestlet partially torn (50–90% intact)
3	Nestlet mostly shredded, often no identifiable nest site, 50–90% shredded, also, less than 50% remains intact, but less than 90% is within a quarter of the cage floor (i.e., not gathered into a nest site but spread throughout cage)
4	An identifiable, but flat nest, more than 90% of the nestlet is torn, the nest is uneven, material is gathered into a nest within a quarter of the cage floor, but the nest is flat with walls higher than mouse body height for less than 50% of its circumference
5	A (near) perfect nest, more than 90% of the nestlet is torn, nest is fairly even, the nest is a crater, with walls higher than the mouse body for more than 50% of its circumference

Source: Deacon [23].

Table 1.
Nesting behaviour rating scale.

2.4.5 Assessment of learning and memory

The Morris water maze developed by Morris [24] for assessing visuospatial learning and memory was used in this study. It was made of a circular polypropylene pool which was divided into four quadrants: Northwest, Northeast, Southwest and Southeast. It measured about 85 cm and 20 cm in diameter and depth respectively. The pool was filled to depth of 14 cm with water. The water was left to sit overnight in order to achieve room temperature (about 26 ± 2°C) and made opaque with the addition of milk to ensure camouflage of the escape platform. The platform was submerged to about 1 cm below the water surface. The pool was located in the laboratory with posters of diagrams hung on the walls to act as visual cues. During testing, the room was dimly lit with diffuse white light. The performance of the animals in the maze was recorded using a camcorder.

Testing in the Morris water maze lasted for 8 days. The first 3 days were for acquisition training with an invisible platform. The next 3 days were for reversal training with the hidden platform in an opposite quadrant. On the seventh day, a probe trial was conducted with no escape platform. On day eight, 4 trials were conducted with a visible platform. Sixty (60) seconds were allocated for each mouse to locate the platform during each trial. Mice which were unable to locate the platform were guided to the position of the platform. The timer was stopped when the mice located the platform within the 60 s. The time it took the mice to locate the platform was recorded as swim latency. After each trial, mice were placed in cages with shredded paper towel beddings to make them dry easily and a heating lamp was also provided to prevent animals from developing hypothermia.

2.5 Statistical analysis

The data derived from the tests were analysed by one-way analysis of variance (ANOVA) followed by post hoc student's Neuma using the computer software SPSS 2007 and Microsoft Excel 2007 for windows vista (Brain Series, China). Data were presented as mean ± SEM (Standard error of the mean) and p value less than 0.05 was considered statistically significant.

3. Results

3.1 Effects of consumption pepper (*Capsicum annum*) and capsaicin on fear and anxiety in the elevated plus maze

The open arm entry frequency of the pepper group was not significantly different compared to the control, whereas that of the capsaicin group was significantly lower ($p < 0.001$) compared to the control (**Figure 1a**). Both the pepper and capsaicin groups had significantly shorter ($p < 0.001$ and $p < 0.01$ respectively) open arm durations compared to the control (**Figure 1b**).

There was no significant difference in the head dips frequency of the pepper-diet fed mice compared to the control (**Figure 1c**). However, that of the capsaicin-diet fed mice was significantly higher compared to the control ($p < 0.01$). While the stretch attends posture of the pepper group was not significantly different compared to control, that of the capsaicin group was significantly higher compared to bot and pepper groups ($p < 0.001$) (**Figure 1d**).

3.2 Effects of consumption of chilli pepper (*Capsicum annum*) and capsaicin on motor coordination in the beam walking test

The foot slips of both the capsaicin and pepper groups were significantly lower compared to the control ($p < 0.001$) (**Figure 2A**). Also, the number of falls of both the capsaicin and pepper groups were significantly lower ($p < 0.001$ and $p < 0.01$ respectively) compared to control (**Figure 2B**).

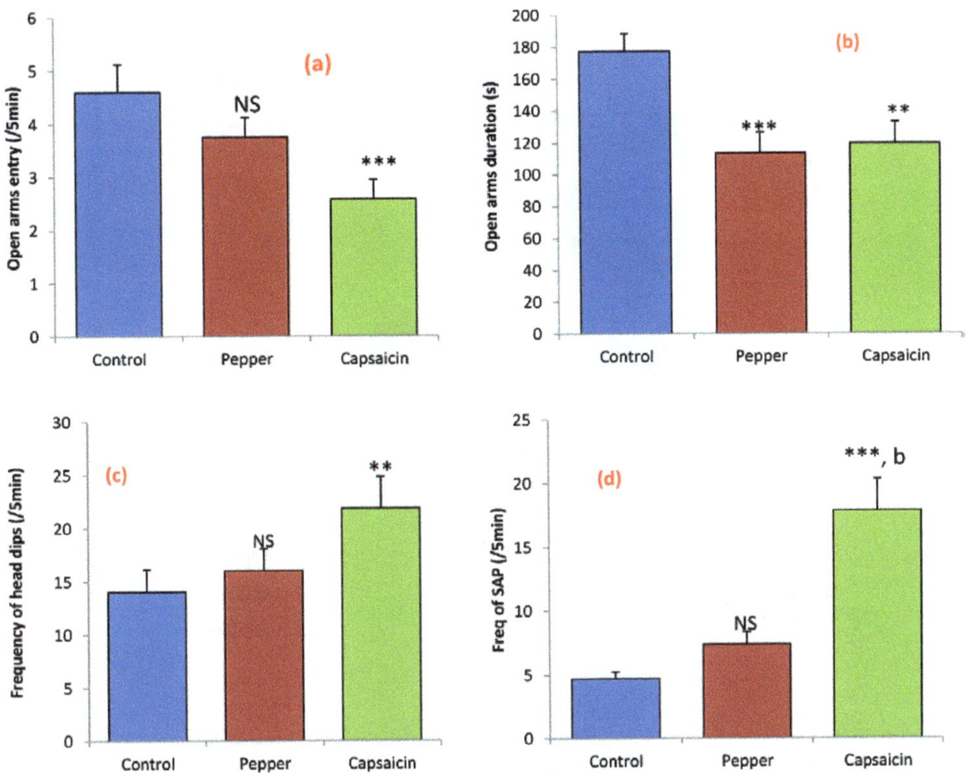

Figure 1.
*Comparison of (a) open arm entry frequency, (b) open arm duration, (c) head dip frequency and (d) stretch attend posture in the elevated plus maze test of the different experimental groups. Values are expressed as mean ± SEM, n = 10. NS = not significant, ** = p < 0.01, *** = p < 0.001 vs control; b = p < 0.001 vs pepper group.*

3.3 Effects of consumption of chilli pepper (*Capsicum annum*) and capsaicin on pain

3.3.1 Tail flick test

During the first test, the latency of tail flick of both the pepper and capsaicin groups were significantly longer ($p < 0.01$ and $p < 0.001$, respectively) compared to control. The value for the capsaicin group was significantly longer ($p < 0.01$) compared to that of pepper group (**Figure 3**).

3.3.2 Hot plate test

The latency of jump of both pepper and capsaicin groups were significantly longer compared to control ($p < 0.01$ and $p < 0.001$, respectively). Meanwhile, the latency of jump of the capsaicin group was significantly longer compared to the pepper group ($p < 0.001$) (**Figure 4a**). While the frequency of hind paw lick of the pepper group was not significantly different compared to the control, the value for capsaicin group was significantly lower compared to both control and pepper groups ($p < 0.001$ and $p < 0.05$, respectively) (**Figure 4b**). Pepper group had a paw lick duration that was not significantly different compared to control. However, the value for capsaicin group was significantly shorter ($p < 0.01$) compared to control (**Figure 4c**).

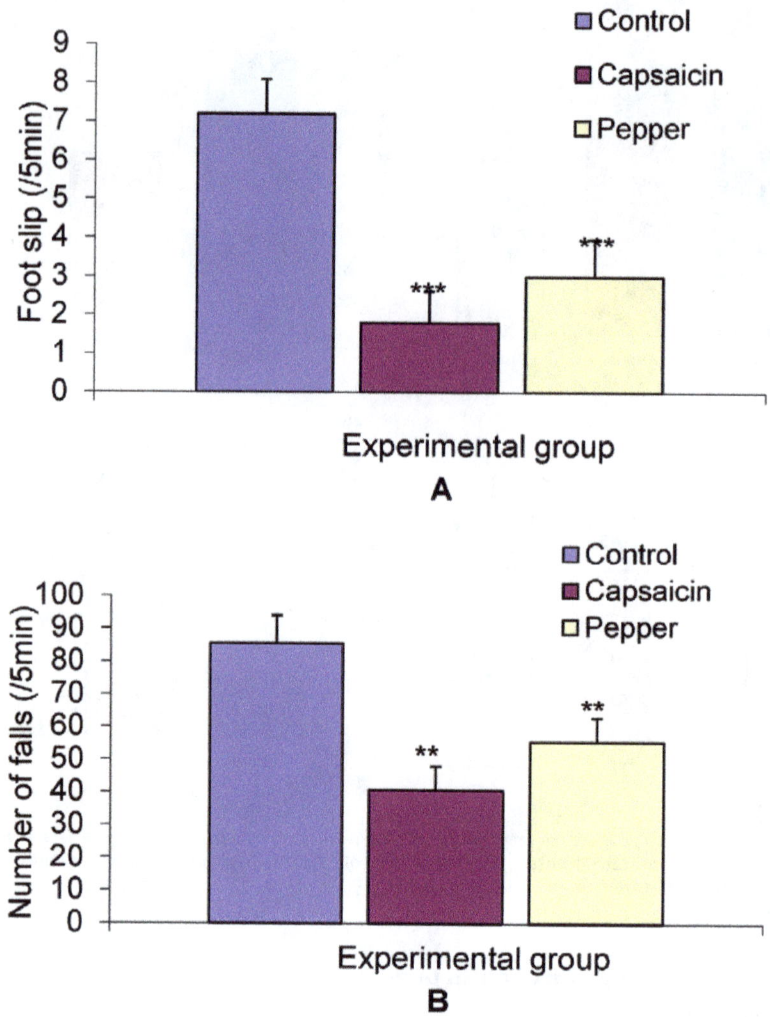

Figure 2.
*Comparison of (A) foot slips and (B) number of falls in the beam walking test of the different experimental groups. Values are expressed as mean ± SEM, n = 10. ** = p < 0.01, *** = p < 0.001 vs control.*

3.4 Effects of consumption of chilli pepper (*Capsicum annum*) and capsaicin on social behaviour in the nesting behaviour test

The nesting score of the capsaicin group was significantly higher ($p < 0.01$) compared to control. The value for the pepper group appeared higher than control but was not significant. However, it was significantly lower ($p < 0.05$) compared to that of the capsaicin group (**Figure 5**).

3.5 Comparison of swim latency in the Morris water maze test for learning

During the acquisition training, the capsaicin group had a significant longer ($p < 0.001$) swim latency on days 2 and 3 compared to control (**Figure 6A**). The swim latency during the reversal training was significantly longer in both capsaicin and pepper groups on day 1 compared to control ($p < 0.001$) but not different on days 2 and 3 (**Figure 6B**).

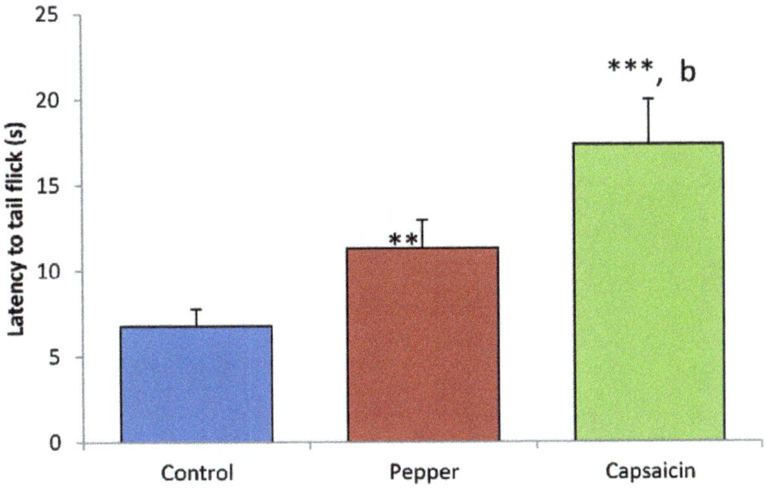

Figure 3.
*Comparison of latency of tail flick in the tail immersion test of the different experimental groups. Values are expressed as mean ± SEM, n = 10. ** = p < 0.01, *** = p < 0.001 vs control; b = p < 0.01 vs pepper group.*

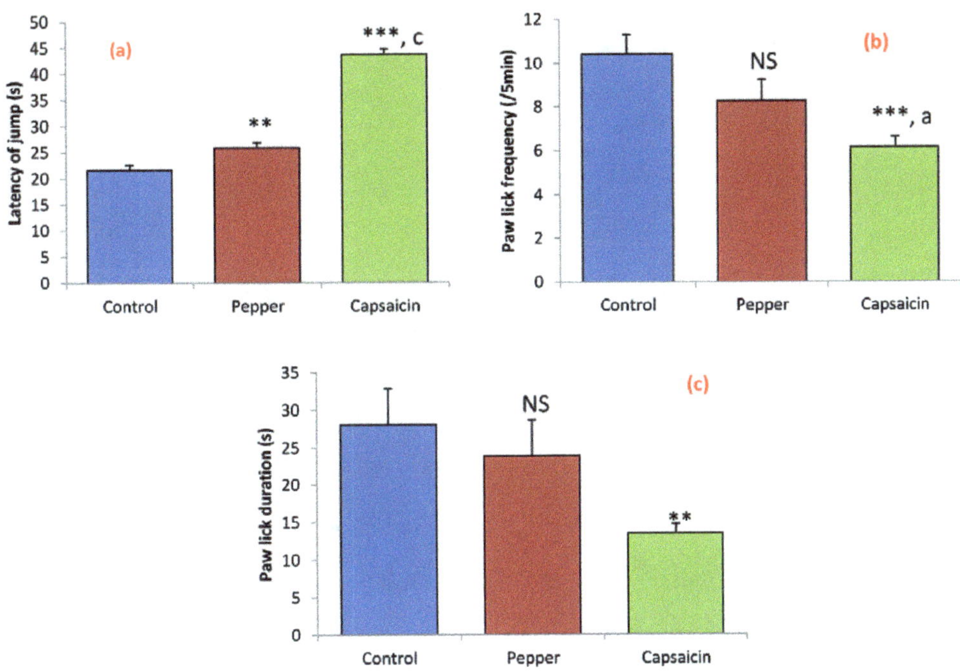

Figure 4.
*Comparison of (a) latency of jump, (b) paw lick frequency and (c) paw lick duration in the hot plate test of the different experimental groups. Values are expressed as mean ± SEM, n = 10. N = not significant, ** = p < 0.01, *** = p < 0.001 vs control; a = p < 0.05, c = p < 0.001 vs pepper group.*

3.6 Comparison of quadrant duration in the Morris water maze test for memory in the different experimental groups

In the probe trial, the pepper and capsaicin groups showed a significantly shorter quadrant duration compared to control ($p < 0.001$ and $p < 0.01$ respectively) (**Figure** 7).

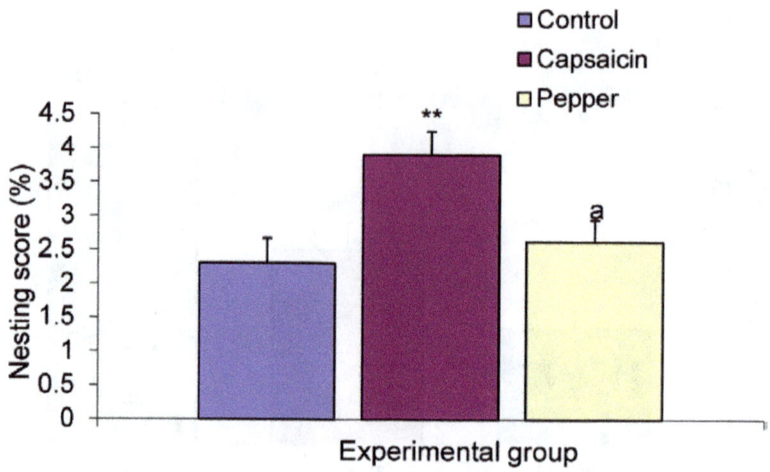

Figure 5.
*Comparison of the nesting score of the different experimental groups. Values are expressed as mean ± SEM, n =10. ** = p < 0.01 vs control; a = p < 0.05 vs capsaicin group.*

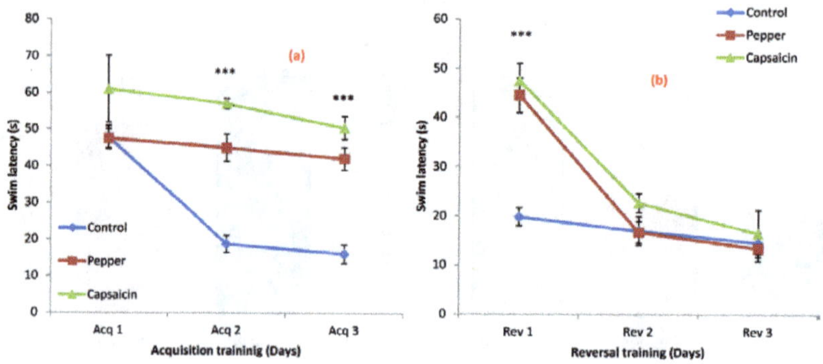

Figure 6.
*Comparison of the swim latencies of the different experimental groups in (a) acquisition and (b) reversal trainings. Values are expressed as mean ± SEM, n = 10. *** = p < 0.001 vs control.*

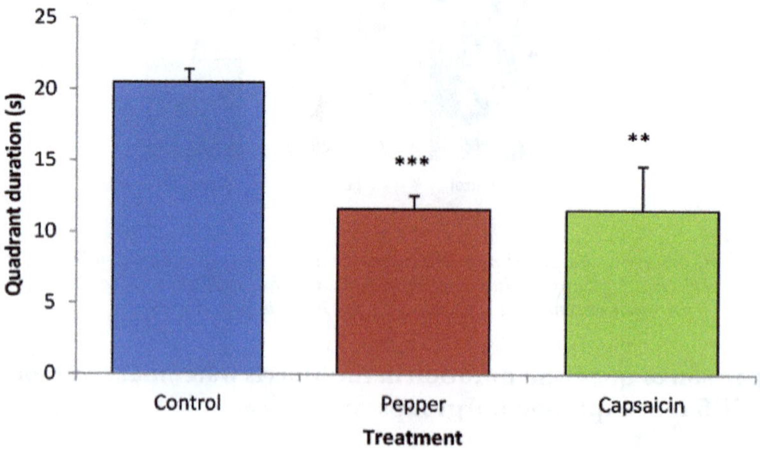

Figure 7.
*Comparison of quadrant duration in Morris water maze test of the different experimental groups Values are expressed mean ± SEM, n = 10. ** = p < 0.01, *** = p < 0.001 vs control.*

4. Discussion

4.1 Anxiety

Following the consumption of pepper and capsaicin diets, the pepper and capsaicin groups had lower open arm entry frequency and duration in the elevated plus maze test. Since fearful mice tend to avoid open areas (especially when they are brightly lit), favouring darker and more enclosed spaces [25], these results imply that chilli pepper and capsaicin caused increase in anxiety in the mice [26].

Increase in risk assessment behaviours such as head dipping and stretch attend posture indicate increased anxiety levels [27]. The results showed that though not significant, the head dips and stretch attend postures of the pepper group were slightly higher than control. On the other hand, the head dips and stretch attend postures of the capsaicin group was significantly higher than the control. These results showed that both pepper and capsaicin increased anxiety in the mice, but the anxiogenic effects of capsaicin were greater [26].

These behaviours confirm the anxiogenic tendencies of long term administration of pepper and capsaicin. However, its mechanism of action has not been ascertained. This is in agreement with the study of Choi et al. [28]. Hakimizadeh et al. [29] also reported that direct injection of capsaicin in the hippocampus induces anxiety-like behaviours, but the report of Santos et al. [30] was on the contrary.

4.2 Motor coordination

Capsaicin and pepper-diet fed mice had significantly reduced foot slips and falls compared to control and these typify improvement in their motor coordina-tion because the lower the frequency of foot slips and number of falls, the better coordinated the animal was [26]. Capsaicin, as the most abundant and commonly occurring capsaicinoid might have achieved this feat (improvement of motor coordination) by aiding in the integration of proprioceptive information with neural processes (TRPV1) in the spinal cord and in the brain (specifically cerebellum) [31].

4.3 Pain

Tail-flick response of a mouse to thermal stimuli is believed to be a spinally mediated reflex behaviour [32]. From the results of our study, the latencies of tail flick of both the pepper and capsaicin-diet fed mice were significantly longer than those of control in the immersion test. Also, the tail-flick latency of the capsaicin group was longer than that of the pepper group. These results imply that pepper or capsaicin diet increased the pain threshold of the mice leading to decreased pain sensitivity. These results point to a more effective analgesic potential in capsaicin than pepper [33].

In the hot plate test, when an animal perceives pain, it attempts to jump away from the object that elicited the painful stimuli. This supra-spinally controlled behaviour (latency of jump) models an escape behaviour. The latencies of jump of both pepper and capsaicin groups were significantly longer than the control. Also, the latency of jump of the capsaicin group was significantly longer than that of pepper group. These results depicted that both capsaicin and pepper exhibited antinociceptive tendencies because it took a long time for mice fed such diets to experience pain compared to control. These results further confirm the analgesic potentials of pepper and capsaicin [33].

Capsaicin may have acted by initially activating heat sensitive TRPV1 recep-tors which induce pain. Repeated and prolonged exposure to capsaicin might have resulted in the reduction of responsiveness of the receptors and ion channels, thus leading to the "defunctionalization" of nociceptor fibres as reported by Anand and Bley [34]. Therapeutic uses of capsaicin for pain treatment were reported; Evangelista [35] and Chung and Campbell [36] reported that capsaicin can be used in treatment of neuropathic pain. Another study reported the analgesic effects of topically applied capsaicin [37]. These reports are in keeping with the results of the present study.

4.4 Social behaviour

The nesting score is an assessment of social behaviour. This nesting behaviour is a reflection of the social behaviour in mice. Nest building in mice correlates to organised behaviour in humans and is very distinct from the findings reported in a research by Alleva et al. [38], where they reported aggressive behaviour in mice treated with capsaicin. A poor performance in the nesting task may indicate impairment in social relationship in mice and likelihood of the presence of autis-tic behaviour. The nesting score of the capsaicin group was significantly higher compared to that of control. The value for pepper group appeared higher but was not significantly higher than the control. This result indicates enhanced organised social behaviour in mice fed with capsaicin diet. However, its mechanism of action has not been ascertained [33].

4.5 Learning and memory

The hidden-platform task of the Morris water maze is a test of visuospatial learning and memory in the mice and is also hippocampus dependent [39]. The use of extra-maze cues was employed in this task. On the other hand, visible-platform (cued) task of the Morris water maze is a non-hippocampal task and dependent on the dorsal striatum (caudate nucleus and putamen) of the basal ganglia [39]. The visible (cued) platform used a unique intramaze visual cue placed at the location of the escape platform.

The shorter the swim latency, the better the training process. Mice with learning disabilities or impairments were not able to quickly figure out the spatial location/ position of the hidden platform, i.e., it took them a long time. Also, the steeper the gradient of swim latencies within the 3 day acquisition or reversal trainings, the better the learning curve, hence faster learning. Following the consumption of pepper and capsaicin diets, the swim latencies of the pepper and capsaicin groups were significantly longer than control in the first 3 days (acquisition training). This shows that pepper and capsaicin delayed learning process during the acquisition training [40].

During reversal training, the swim latencies of the test groups were also significantly longer on day 1 of the 3 day reversal training task, while on days 2 and 3, the values did not differ from control. This means that on days 2 and 3, the three groups learned equally while the control learned better on day 1. Visuospatial memory was also assessed during the probe trial (exploration without hidden platform). During this trial, it was expected that mice with good memory of the spatial location/posi-tion of the hidden platform would spend more time exploring the quadrant which had the platform during reversal training (North-East quadrant), but this was not observed in mice treated with pepper and capsaicin diets. They spent less time in the North-East quadrant. This means that they had memory impairment [40]. This is in contrast to the work by Kong et al. [41] which reported that capsaicin did not significantly alter the learning and memory performance in young adult mice but reduced the number of newly generated cells in the hippocampus. However, this is in line with the work by Kooshki et al. [42]. It is possible that the nociceptive effects of Capsaicin might have also affected learning and memory in the mice.

Learning and memory which are complex cognitive functions of the higher nervous centres encompass a variety of subcomponents with many interactions and overlaps [43]. Memories are stored in the brain by changing the basic sensitivity of synaptic transmission between neurons as a result of previous neural activity. The effects observed might have been due to the presence of the alkaloid called capsaicin. Since capsaicin is neurotoxic [9], it is likely that it impaired synaptic transmission between neurons by interfering with the basic sensitivity of the transmission in the hippocampus leading to impairment of learning and memory of the mice.

5. Conclusion

Consumption of both chilli pepper and capsaicin diets decreased pain perception, increased anxiety related behaviours, impaired learning and memory but enhanced social behaviour and motor coordination in mice. Therefore, it is likely that capsaicin, which is a powerful and stable alkaloid in chillies, may be one of the principles responsible for the observed neurobehavioural parameters in the experimental animals.

Author details

Eme Efiom Osim[1,2*], Anyauba Nmaju[2], Azubuike Nwankwo[2], Udemeobong Okon[1] and Joshua Iwasam[1]

1 Department of Physiology, University of Calabar, Calabar, Nigeria

2 Department of Physiology, Abia State University, Uturu, Nigeria

*Address all correspondence to: emeosim@yahoo.com

References

[1] Owolabi. Identified Capsicum Annum, Assigned a Voucher Number and Deposited it in the Herbarium of Botany Department of University of Calabar. Calabar, Nigeria; 2015

[2] Yoon JY, Green SK, Tshanz AJ, Tsou SCS, Chenge LC. Pepper improvement for the tropics, problems and approach. In: Tomato and Pepper Production in the Tropics. Asian Vegetable Research and Development Center. Taiwan: Shantung; 1989. pp. 86-90

[3] Zimmer AR, Leonardi B, Zimmer ER, Kalinine E, de Souza DO, Portela LV, et al. Long-term oral administration of *Capsicum baccatum* extracts does not alter behavioral, hematological and metabolic parameters in CF1 mice. Evidence-based Complementary Alternative Medicine. 2012;**2012**:196358

[4] Katzer G. Paprika (*Capsicum annum* L.). 2008. Available from: https://www.unigraz.at/~katzer/engl/caps_ann.html [Accessed: 20 October 2017]

[5] Johnson W. Final report on the safety assessment of *Capsicum annuum* extract, *Capsicum annuum* fruit extract, *Capsicum annuum* resin, *Capsicum annuum* fruit powder, *Capsicum frutescens* fruit, *Capsicum frutescens* fruit extract, *Capsicum frutescens* resin, and capsaicin. International Journal of Toxicology. 2007;**26**(1):3-106

[6] Nagy Z, Westerberg H, Klingberg T. Maturation of white matter is associated with the development of cognitive functions during childhood. European Journal of Pharmacology. 2004;**500**:351-369

[7] Yang F, Zheng J. Understanding spiciness: Mechanism of TRPV1 channel activation by capsaicin. Protein & Cell. 2017;**8**(3):169177

[8] Basith S, Cui M, Hong S, Choi S. Harnessing the therapeutic potential of capsaicin and its analogues in pain and other diseases. Molecules. 2016;**21**(8):966

[9] Kempaiah RK, Manjunatha H, Srinivasan K. Protective effect of dietary capsaicin on induced oxidation of low-density lipoprotein in rats. Journal of Molecular and Cellular Biochemistry. 2005;**275**:7-13

[10] Deal CL, Schnitzer TJ, Lipstein E, Seibold JR, Stevens RM, Levy MD, et al. Treatment of arthritis with tropical capsaicin: A double blind-trial. Clinical Therapeutics. 1991;**13**(3):383-395

[11] Holst H, Arendt-Nielsen L, Mosbech H, Serup J, Elberling J. Capsaicin-induced neurogenic inflammation in the skin in patients with symptoms induced by odorous chemicals. Skin Research and Technology. 2011;**17**(1):82-90

[12] Corey JH. Riot control agents. In: Gupta RC, editor. Handbook of Toxicology of Chemical Warfare Agents. 2nd ed. New York: Elsevier; 2015. pp. 131-150

[13] Szolcsányi J. Capsaicin and sensory neurons: A historical perspective. Progress in Drug Research. 2014;**68**:1-37

[14] Lister RG. The use of a plus maze to measure anxiety in the mouse. Psychopharmacology. 1987;**92**(2):180-185

[15] Okon UE, Erigbali PP, Osim EE. Comparative effects of antiepileptic agents Dichrostachys glomerata ethanol extract and carbamazepine on seizures and anxiety in mice. Journal of Advances in Medicine and Medical Research. 2017;**24**(4):1-10

[16] Trullas R, Skolnick P. Differences in fear motivated behaviours among inbred mouse strains. Psychopharmacology. 1993;**111**:323-331

[17] Curzon P, Zhang M, Radek RJ, Fox GB. The behavioural assessment of sensorimotor processes in the mouse: Acoustic startle, sensory gating, locomotor activity, rotarod and beam walking. In: Buccafusco JJ, editor. Methods of Behavior Analysis in Neuroscience. 2nd ed. Florida: Taylor & Francis Group; 2009

[18] Ramabadran K, Bansinath M, Turndorf H, Puig MM. Tail immersion test for the evaluation of a nociceptive reaction in mice, methodological considerations. Journal of Pharmacological Methods. 1989;**21**(1):21-31

[19] Huber JD, Campos CR, Egleton RD, Witt K, Guo L, Roberts MJ, et al. Conjugation of low molecular weight poly (ethylene glycol) to biphalin enhances antinociceptive profile. Journal of Pharmaceutical Sciences. 2003;**92**(7):1377-1385

[20] Le Bars D, Gozairu M, Cadden SW. Animal models of nociception. Pharmacological Reviews. 2001;**53**(4):597-652

[21] Bertollo CM, Oliveira AC, Rochas LT, Costa KA, Nascimento EB, Coelho MM Jr. Characterization of the anti-nociceptive and anti-inflammatory activities of riboflavin in different experimental models. European Journal of Pharmacology. 2006;**547**(1/3):184-191

[22] Bender BS, Cottey R, Bell W, Taylor S. Body temperature and nesting behaviour following influenza challenge in mice: Effects of age. Mechanisms of Ageing and Development. 1996;**86**(1):1-9

[23] Deacon RM. Assessing nest building in mice. Nature Protocols. 2006;**1**(3):1117-1119

[24] Morris R. Developments of a water maze procedure for studying spatial learning in the rat. Journal of Neuroscience Methods. 1984;**11**:47-60

[25] Bailey KR, Crawley JN. Anxiety-related behaviors in mice. In: Buccafusco JJ, editor. Methods of Behaviour Analysis in Neuroscience. 2nd ed. Florida: Taylor & Francis Group; 2009

[26] Nmaju AU, Joshua IE, Okon UE, Bisong SA, Nwankwo A, Osim EE. Long-term consumption of *Capsicum annuum* (chili pepper) and capsaicin diets elevates anxiety but improves motor coordination in CD-1 mice. Journal of Advances in Medicine and Medical Research. 2018;**25**(9):1-10

[27] Holmes A, Kinney JW, Wrenn CC, Li Q , Yang RJ, Ma L, et al. Galanin GAL-RI receptor null mutant mice display increased anxiety-like behaviour specific to the elevated plus maze. Neuropsychopharmacology. 2003;**28**(6):1031-1044

[28] Choi YJ, Kim JY, Yoo SB, Lee JH, Jahng JW. Repeated oral administration of capsaicin increases anxiety-like behaviours with prolonged stress-response in rats. Journal of Biosciences. 2013;**38**(3):561-571

[29] Hakimizadeh E, Oryan S, Hajizadeh-Moghaddam A, Shamsizadeh A, Roohbakhsh A. Endocannabinoid system and TRPV1 receptors in the dorsal hippocampus of the rats modulate anxietylike behaviours. Iranian Journal of Basic Medical Sciences. 2012;**15**(3):795-802

[30] Santos CJ, Stern CA, Bertoglio LJ. Attenuation of anxiety-related behaviour after the antagonism of transient receptor potential

vanilloid type 1 channels in the rat ventral hippocampus. Behavioural Pharmacology. 2008;**19**(4):357-360

[31] Aman JE, Elangovan N, Yeh IL, Konczak J. The effectiveness of proprioceptive training for improving motor function: A systematic review. Frontiers in Human Neuroscience. 2014;**8**:1075

[32] Ito S, Okuda-Ashitaka E, Minami T. Central and peripheral roles of prostaglandins in pain and their interactions with novel neuropeptides nociceptin and nocistatin. Neuroscience Research. 2001;**41**(4):229-332

[33] Nmaju AU, Joshua IE, Okon UE, Nwankwo AA, Osim EE. Long-term consumption of *Capsicum annum* (chili pepper) and capsaicin diets suppresses pain perception and improves social behavior of CD-1 mice. Nutrition and Food Science. 2018;**48**(6):911-921

[34] Anand P, Bley K. Topical capsaicin for pain management: Therapeutic potential and mechanisms of action of the new high-concentration capsaicin 8% patch. British Journal of Anaesthesia. 2011;**107**(4):490-502

[35] Evangelista S. Novel therapeutics in the field of capsaicin and pain. Expert Review of Clinical Pharmacology. 2015;**8**(4):373-375

[36] Chung MK, Campbell JN. Use of capsaicin to treat pain: Mechanistic and therapeutic considerations. Pharmaceuticals (Basel). 2016;**9**(4):66

[37] O'Neill J, Brock C, Olesen AE, Andresen T, Nilsson M, Dickenson AH. Unravelling the mystery of capsaicin: A tool to understand and treat pain. Pharmacological Reviews. 2012;**64**(4):939-971

[38] Alleva E, Aloe L, Bigi De Acetis L. Capsaicin affects aggressive behaviour, but not hot plate responding, of adult male mice. Physiology & Behavior. 1991;**49**(4):715-719

[39] McDonald RJ, White NM. Parallel information processing in the water maze: Evidence for independent memory systems involving dorsal striatum and hippocampus. Behavioural Neural Biology. 1994;**61**:260-270

[40] Nmaju AU, Joshua IE, Okon UE, Nwankwo AA, Osim EE. Long-term consumption of *Capsicum annum* (chili pepper) and capsaicin diets impairs Visuo spatial learning and memory in CD-1 mice. Journal of Advances in Medicine and Medical Research. 2017;**24**(7):1-12

[41] Kong KH, Kim HK, Song KS, Woo YS, Choi WS, Park HR, et al. Capsaicin impairs proliferation of neural progenitor cells and hippocampal neurogenesis in young mice. Journal of Toxicology and Environmental Health. 2010;**73**(21-22):1490-1501

[42] Kooshki R, Abbasnejad M, Esimaeili-Mahani S, Raoof M. The role of trigeminal nucleus caudalis orexin 1 receptors in orofacial pain transmission and in orofacial pain-induced learning and memory impairment in rats. Physiology & Behavior. 2016;**157**:20-27

[43] Brem A, Ran K, Pascual-Leone A. Learning and memory. Handbook of Clinical Neurology. 2013;**116**:693-737

9

Use of Capsaicin for Nonlethal Technology

Nilton Oliveira Junior, Valter Padulla, Vinnicius Ferraço and Giovana Ferronato

Abstract

Hot peppers have been used as a seasoning for food, because of both the characteristic flavor and the peculiar feature of pungent sensation. In the same way, it has been used as cosmetics and in other forms such as topical stimulants and body creams. Nonetheless, this chapter shows a not-so-common application of pepper's properties, like the so-called pepper spray for defense technology. According to the premise that the action of the police officer must be adapted to each situation, it is necessary to prepare him for the gradual and proportional use of force, avoid-ing excesses and even lethal resources. In this way, pepper spray is a resource that can place police action within this premise. The main aspects of pepper spraying technology are discussed and the main concepts related to use, hazards, solution formulations and capsaicinoids quantification tests are pointed out. The advance showed here is around the use of capsaicinoids to extract raw material instead of OC to produce pepper spray solutions. The absence of oils and resin is the main point, as nonflammability is achieved much easier and avoids the use of hazardous organic solvents. The chemical analyses by GC-MS/FID applied to quantify the capsaicinoids, with capsaicin and dihydrocapsaicin together, are demonstrated and can be reproducible for the quality control of this product.

Keywords: pepper spray, nonlethal technology, capsaicinoids, capsaicin

1. Introduction

The Article XII in the Declaration of the Rights of Man and of the Citizen—France—1789, declares that "*The guarantee of the rights of man and of the citizen necessitates a public force; this force is thus instituted for the advantage of all and not for the particular utility of those in whom it is trusted*".[1] The exercise of police work in civic security is very complex and thorough; it requires, at the same time, the application of the law to those who do not want to comply with it and requires the correct, just and necessary application of force to maintain law and order. If, on the one hand, the violation of the law is not correct, the same law provides that this citizen must be treated by the State according to the citizen protective premises of that law [1].

The State, through the Police, has to act with legality and attention to the protec-tion of the human person; in other words, "police solutions" to violations of the law

[1] Declaration of the Rights of Man and of the Citizen—France—1791.

must not incur excesses by law enforcement officers. In order to adjust the conduct of law enforcement officers, there is a need to prepare to use force gradually, according to the premise that police action should suit each situation—considering that most police cases do not involve the use of lethal weapons—so, the police response must include nonlethal solutions [2].

The gradual use of force is an internationally recognized concept, which aims to point out the best way, in terms of containment, during police assistance [3]. Resisting the use of nonlethal technologies means exposing public security professionals to irreparable excesses and errors, due to their effects.

Among the nonlethal solutions for the use of force (in response to police incidents) are disabling agents that do not require physical contact between the police and the lawbreaker (for example, pepper spray), with the advantage of maintaining the offender at a safe distance and unable to harm others, including himself and the police officer.

We must be aware of the growing issue of police officers who are injured during police duties. There is great public attention, fully justified and correct, regard-ing the excessive actions that can result from the confrontation between citizens and police officers, with a primary focus on avoiding lethality in these situations. However, it is important to understand that the State, through its agents, vulnerability is also not desirable, that is, the impediment to the use of lethal actions could be compensated by the access to nonlethal solutions, more immediate equipment, as previously said, the pepper spray is an example.

Still with respect to the vulnerability of the police, there is the attack of animals (usually dogs), against police officers or third parties; it also strongly indicates the use of pepper spray as a nonlethal solution. There are reports of criminals who train dogs to attack police officers, or even police dogs; such a condition would be mitigated with the use of pepper spray.

2. Pepper spray

Usually, hot peppers are used as a seasoning for food, because of both the characteristic flavor and the peculiar feature of pungent sensation. In the same traditional way, they have also been used as cosmetics and in other forms such as topical stimulants and body creams. Nonetheless, this chapter shows a not-so-common application of hot pepper's properties, like the so-called pepper spray for defense technology. Since the 1970s, pepper spray is used as a nonlethal defense solution by law enforcement forces and civilians around the world [4].

The irritating effect—when the solution contained in the spray is sprayed on the face—allows for disabling the opponent by providing some advantage in self-defense (or third party), crowd controlling, and dispersing civilian disturbances. The immediate and involuntary closure of the eyelids and lacrimation were the main expected effects. Burning eyes, cough, nasal discharge, difficulty breathing, burning in the mouth, and other reactions occur within seconds of exposure. Those pain effects, together with the psychological (or moral) reflexes, cause temporary disability, thereby allowing police officers to avoid the opponent's resistance or even counteracts. The temporary incapacitation lasts for about 10–15 min and, after a few hours, the individual self-regains full capacity without any permanent effect, in other words, without irreversible harm. These properties take the pepper spray for defense as a nonlethal defense technology, an alternative to lethal force.

There are cases of serious injuries from the use of pepper spray, like the situations reported were the mechanical damage to the eye by the high-pressure aerosol jet applied directly to the eye [5], but this can be avoided through training and respecting the safe distance for application. Use in closed spaces, or "in door" uses, is dangerous too, because the spray can asphyxiate in closed spaces, for no other reason other than the aerosol taking the place of breathable air. Several authors have studied the permanent injury cases with the use of pepper sprays (when used improperly), and the most important permanent injury is cornea damage, followed by pulmonary injury cases [6–10]. Nevertheless, in general, the reported cases of death always involve adverse circumstances, due to either the environment or the individual, or even both, which leads to death for reasons other than the single effect of the irritant. The inverse problem is dangerous too, because the inefficiency of the incapacitating properties of the spray solution leads the law agents at risk [11]. Both injuries and inefficiency should be considered when developing pep-per spray solutions. The challenge is to find the best formulation that is efficient without causing damage.

Although there are several types of hot pepper extracts, the most widely used irritant agent in the production of pepper sprays is the oleoresin capsicum (OC) [12]. OC is a viscous liquid extracted from the hot peppers fruit (*Capsicum*), which contains more than a hundred different types of substances, consisting of complex mixtures, mainly of lipophilic molecules, and does not have a simple chemical formula [13].

Oleoresin can be just a colorant red type (paprika), colorant and pungency type (red pepper), or high pungent type (capsicum). With regard to industrial production, the capsicum OC pepper spray type has been the most supplied to police forces.

The nonivamide (pelargonic acid vanillylamide or PAVA) is a surrogate substance and some manufacturers use it instead of OC or capsicum extracts. This irritant can be found in some pepper extracts at lower concentrations; however, PAVA can be synthetically made, at low costs, and used alone or together with OC in some formulations. PAVA is much less irritating and its effect is felt much later than capsaicin.

Notwithstanding, there are formulation improvements by the application of another irritant capsaicin content that presents more homogeneous composition and can be more effortlessly controlled than OC. This irritant is an alternative type of pepper extract named natural capsaicin (N.Cap) [14] that engenders high-quality sprays. In practice, the irritant effect is not so different from OC but have a better-defined chemical composition and other advantages. The first advantage point is the absence of red dyes plus the characteristic seasoning smell of the OC, and the second point is the absence of oils and resins making possible the totally nonflammable solution formulation. Therefore, the N.Cap spray solutions cause just pungent effects and nothing else.

The N.Cap can be purchased from manufacturers that supply common pepper extracts to the pharmaceutical market. The product presents a white crystalline coarse powder with the presence of flat-shaped flakes. The irritant content requirement meets a minimum of 95% capsaicin plus dihydrocapsaicin. The remaining 5% consists of the extraction residues at pharmaceutical levels and by other capsaicinoids. Furthermore, it meets maximum drying loss of 1%, maximum ignition residue of 1%, and residual solvent (methanol) max 3000 ppm (USP-467). It has a melting range of 57–66°C. Metal content is <10.0 ppm and arsenic <2.0 ppm.

3. The chemical composition of OC

The chemical composition of OC depends on several factors—the species of plant, the extraction method, the season of the year in which the crop is harvested, and the region of the plant where it is cultivated—as well as processing way and production. All of these and a lot of other factors contribute to make it difficult to standardize the concentration of irritating agent in pepper spray solutions, thereby leading to a wide range of capsaicin amounts contained in different commercial products and also in separate lots from the same supplier.

Since they produce the desired irritant effect, the capsaicinoids are the most important substances present in the OC used in pepper spray solutions. Other substances such as phenols, acids, alcohols, aldehydes, carotenes, esters, oils, and resins are also present in the OC, but cause a little or no irritant effect compared to capsaicinoids. It should be noted that oils and resin compounds are potentially flammable, and they become undesirable by increasing the flammability of the spray solution. If the spray solution has flammable properties, a jet of flame can be produced if there is an ignition source in the path of the spray jet.

Furthermore, the OC normally has a strong spice smell, a striking reddish brown color, and has hydrophobic characteristics, thereby requiring organic solvents (or even emulsions) for its complete dissolution. All these features mentioned are undesirable in pepper spray technology.

The OC extraction has wide types of processes, and the most simple is the extraction through ground powder fruits. Organic solvents like hexane, ether, alcohols, acetates, and ethylene dichloride are common extraction liquid phase media. In general, the extractions pass by two or three times, removing and replacing with pure solvent at each time. Other simple extractions consist basically to macerate the fruit with olive oil or other seed oils, but the impurity is high and the standardiza-tion is far from ideal to use as a raw material for pepper sprays. Modern processes of extractions are advanced Soxhlet extraction, microwave-assisted extraction, supercritical fluid extraction, ultrasound-assisted extraction, and pressurized liquid extraction. All they are more suitable for better-controlled extractions. However, these methods are very expensive and greatly increase the production costs [13].

The concentration of capsaicinoids in OC, in general, shows variations from 1 to 10%, depending on the type of fruit, solvent, process, and so on. Some types of OC capsicum have an approximate concentration range of 3–6% of capsaicinoids, but some Indian suppliers may have OC with 20% of capsaicinoids [15]. Of course, this large amount of concentrations and side substances present in the OC leads to a difficult standardization of raw material for the pepper spray proposal [12].

4. Capsaicinoids

The substances so-called capsaicinoids can be represented, in a simple way, by the binding of a vanillylamide group with fatty acids as a principal characteristic molecular structure, where the fatty chain would have 9–11 carbon atoms. The capsicum fruits synthetizes the capsaicinoids in the placenta-fruit region, close to the tissue adjacent to the seeds. This type of metabolite is unique to *Capsicum* genus. Although little data exist on this compound's biosynthesis, there are reports that the principal two pathways are phenylpropanoid and fatty acid metabolism [16]. The amount of capsaicinoids in the pepper fruit reaches the maximum concentration in about 2 or 3 months of natural fruit development, and after the deceleration initiated by the action of the peroxidase enzymes, the concentration decreases generating secondary compounds. Certainly, the concentration in the fruit varies with the plant species, cultivation conditions, soil, place, seasonality, and so on.

From the point of irritant effect, the capsaicin is the main substance in the midst of capsaicinoids compounds, followed by dihydrocapsaicin. The total capsaicinoids content in the fruits has a typical concentration range of 0.1–1.0% on a dry basis [17], but it is not a rule and some species have lowest concentration like 0.003% and others have a far high concentration like 1.86% [18]. In general, the capsaicin and dihydrocapsaicin together correspond to ~90% of the total capsaicinoids present in the ready fruit. The typical distribution is 60–70% capsaicin, 20–30% dihydrocapsaicin, 3–7% nordihydrocapsaicin, ~1% homocapsaicin, and ~ 1% homodihydrocapsaicin. However, this is not true in all cases; once, for example, there is evidence that the degree of fruit maturation and incidence of sunlight in the cultivation area, temperatures, and water availability are important factors in the accumulation and proportion of capsaicinoids in the fruit [19]. **Table 1** shows the name, chemical structure, and molecular formula of the major capsaicinoids.

Unfortunately, there are associations between OC and the molecular capsaicin formula in some pepper spray technicians media, thereby leading to erroneous perception that OC and capsaicin are synonymous with the same substance. Nevertheless, the molecular formula $C_{18}H_{27}NO_3$ is not the OC formula but only the capsaicin molecular formula, which is contained in OC.

4.1 The biological action of capsaicinoids

The biological action of capsaicinoids is complex and multi-targeted mechanisms. The action involves, above all, the activation of peripheral nerve receptors

Homolog name	Chemical structure	Molecular formula	CAS
Capsaicin	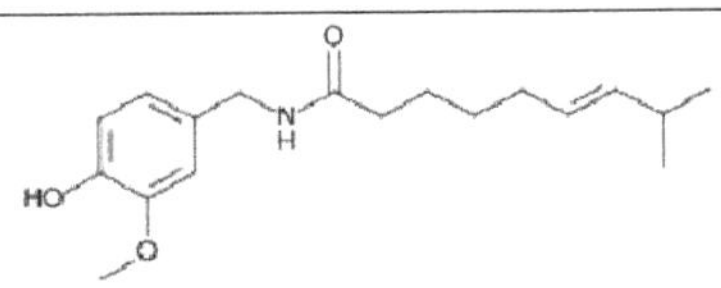	$C_{18}H_{27}NO_3$	404-86-4
Dihydrocapsaicin	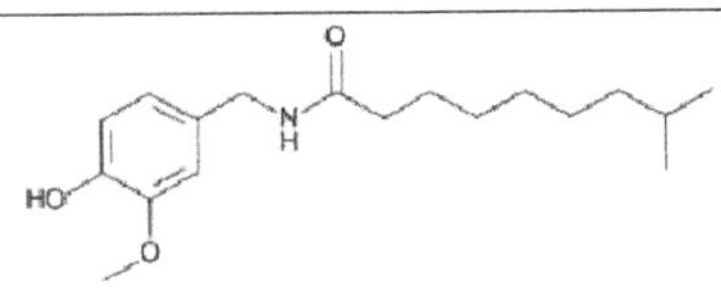	$C_{18}H_{29}NO_3$	19408-84-5
Nordihydrocapsaicin	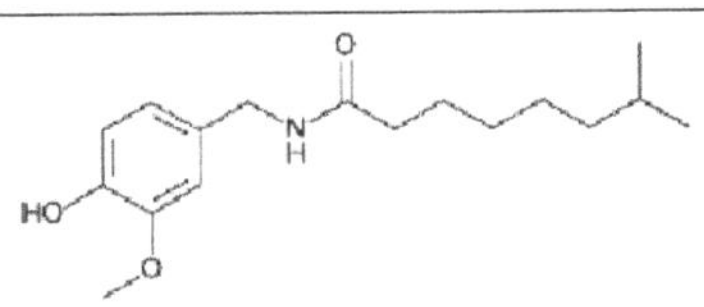	$C_{17}H_{27}NO_3$	28789-35-7
Homocapsaicin	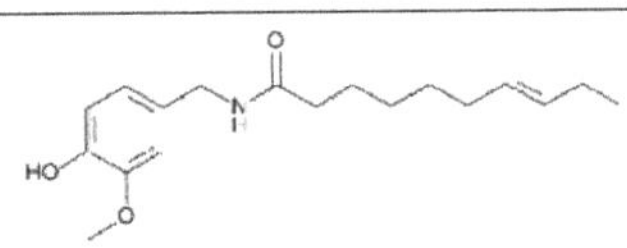	$C_{19}H_{29}NO_3$	58493-48-4
Homodihydrocapsaicin	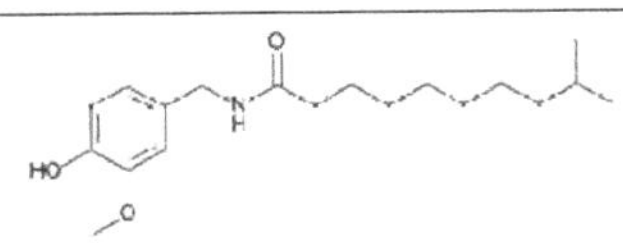	$C_{19}H_{31}NO_3$	20279-06-5
Nonivamide (PAVA)	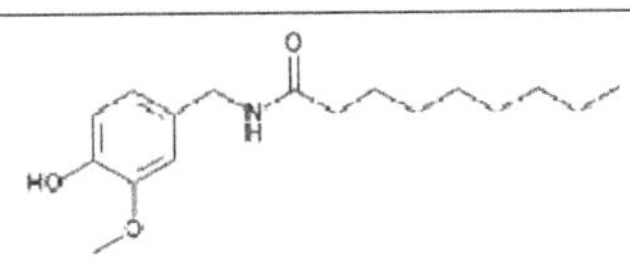	$C_{17}H_{27}NO_3$	2444-46-4

Table 1.
Name, chemical structure, molecular formula, and CAS number of the major capsaicinoids.

in the mucous membranes, by interactions of the capsaicinoids with the vanilloid transient receptor potential type-1 cation channels (TRPV1). These receptors are normally activated by temperatures between 37 and 45°C when opening the calcium channels and induce reflexes of burn sensation. However, when TRPV1 binds with the capsaicinoids, it causes these channels to open below 37°C and the burn sensa-tion occurs at normal body temperature. This is why capsaicinoids are linked to the sensation of heat [20].

Furthermore, the interactions of the TRPV1 with capsaicin are strictly related to pain by nociceptor activation and the release of substance-P. Nociceptor is a nervous sensory receptor responsible for pain mechanisms in the human body [21]. Prolonged contact may cause nerve endings to be desensitized; however, it does not lead to a permanent desensitization state and can be reversed by discontinuing contact.

The capsaicin and dihydrocapsaicin content in the pepper spray solution is determinant for the irritant properties and the disabling effect of the spray. Thus, to evaluate the effectiveness of the disabling properties of the spray, the capsaicinoids concentration in the solution can be measured, particularly the capsaicin together with dihydrocapsaicin [14]. Obviously, the limits to this concentration must take into account the toxicity and security against injury. So, some connections among the concentration and the irritant effect must be evaluated.

The Scoville test, whose unit of measure is SHU, is a known form of evaluating the effect of pungency or blazing of the peppers. This method was developed in 1912 by Wilbor Scoville [22] giving a five-level scale for pungency: nonpungent (0–700 SHU), low pungency (700–3000 SHU), moderate pungency (3000–25,000 SHU), high pungency (25,000–70,000 SHU), and very high pungency (>80,000 SHU). However, the Scoville method is a taste organoleptic method made by a dilution series of the pepper extracts. It makes subjective responses and unreproducible results. Qualitative and quantitative information can be precisely obtained by modern instrumental chemical methods with many advantages [12, 23].

5. Toxicology of capsaicin

Since the peppers are largely used as a food, through several forms and quantities around the word for centuries, it is not expected that it is a poison. The individual consumption of capsaicin in India may be around 7–120 mg/day. In Northeast Thailand, the individual consumption may be 26 g of (jaew) pepper per meal [24]. The main concern with the use of pepper spray is with the possible toxic properties of the solution. Another important aspect is the observation of the tactical way of use, which involves the amount of solution applied, environment conditions, closed spaces and psychological state of the aggressor; all of these characteristics must be well studied in action using pepper spray.

Previous study shows that the toxic level of 60-kg human consumption of capsaicin corresponds to consumption of 1.94 kg of dry weight of capsicum fruit. Obviously, there is no person who can consume this amount of dry pepper at once; due to the pain and pungent sensation, it prevents over consumption [25].

There is no consensus in the literature for the acceptable dose of capsaicin for oral, skin, and eyes human contact. The studies are commonly connected with lethal doses values (LD_{50}), which means a quantity limit that kills 50%, at least, of the population studied. This kind of test programs normally uses rats or mouse to access results. Rabbits, dogs, and guinea pig are common too. It is sustained by a presumption that these animals have a close correlation with human responses to capsaicin [26].

An oral LD_{50} value was reported at range of 60–75 mg/kg (Swiss male albino mice). On the other hand, another experiment with the same mice type and procedure, just changing the solution vehicle, showed LD_{50} at 190 mg/kg [25]. Another work reports male mice with LD_{50} values of 118.8 mg/kg and for female mice of 97.4 mg/kg [27]. The U.S. National Library of Medicine—National Center for Biotechnology Information—reports 47.2 mg/kg for mouse. OSHA or NIOSH has reported no occupational exposure limits for capsaicin or OC. As can be seen in these short examples above, there is not a single value to use to computation and derive secure concentrations of spray solutions to manufactories.

Some inhalation response studies showed no evidences that inhalation of capsicum oleoresin spray causes respiratory compromise [28]. An investigation made with the concern on the respiratory effects of OC concludes that the exposure did not result in abnormal hypoxemia or hypoventilation. This experiment was made by aerosol delivery exposure box with a hood for the subject attached to one end of the exposure box. The aerosol was 5.5% OC (with 0.92% of capsaicinoids) solution with isopropyl alcohol as carrier agent and isobutane/propane as propellant [29]. These types of studs are more conclusive and can take better access for more suitable results to manufactories.

The ocular contact with capsaicin is the primary incapacitation response. Furthermore, the hydrophobic properties of capsaicin (and capsaicinoids) allow them to penetrate the eye tissue, accessing the terminal nervous and, consequently, causing pain and great lacrimation response. The use of contact lenses can lead to an increase in the duration of the effect, due to the accumulation under the lens. In these cases, care should be taken to remove the lens as soon as possible.

6. Experimental

The quantification of capsaicin in pepper spray can be made by GC-MS/FID [14]. This experimental result presented here aims to exemplify the quantitative and qualitative analysis of the samples of the commercial Brazilian pepper spray. About 500 µl of solution sample was weighed on an analytical balance and the contents were solubilized with HPLC grade acetone and transferred quantitatively to a 1 ml volumetric flask. The solution was swollen and an aliquot was conducted for analysis by gas chromatography.

A model GC-2010 equipped with a mass spectrometer GCMS-QP2010 Ultra and an automatic sampler AOC-5000 (Shimadzu) was used. The chromatographic separation was performed using RTx-5MS capillary column (Restek) with a stationary phase containing 5% phenylmethyl and 95% polydimethylsiloxane (30 m × 0.25 mm × 0.25 µm of film). The temperatures of the injector and the transfer line of the mass spectrometer were 300°C. The samples were injected with a split ratio of 1:50. The oven temperature program followed the following condi-tions: 50°C (2 min), heating rate from 10°C/min to 300°C (2 min). Helium was used as the carrier gas in a flow rate of 1.2 ml/min. The mass spectrometer was operated in electron impact ionization mode, with acquisition in the scan mode with m/z between 50 and 500.

To identify the compounds, a comparison of the spectra with the NIST library was carried out. Only substances whose similarity was greater than 80% were considered, compared with data from the library. **Figure 1** shows the chromatogram result with the solvent around 7.2 min and capsaicin and dihydrocapsaicin peak around 24 min.

With the purpose of quantifying, the capsaicin and dihydrocapsaicin are integrated together and the calibration curve was made by this result against

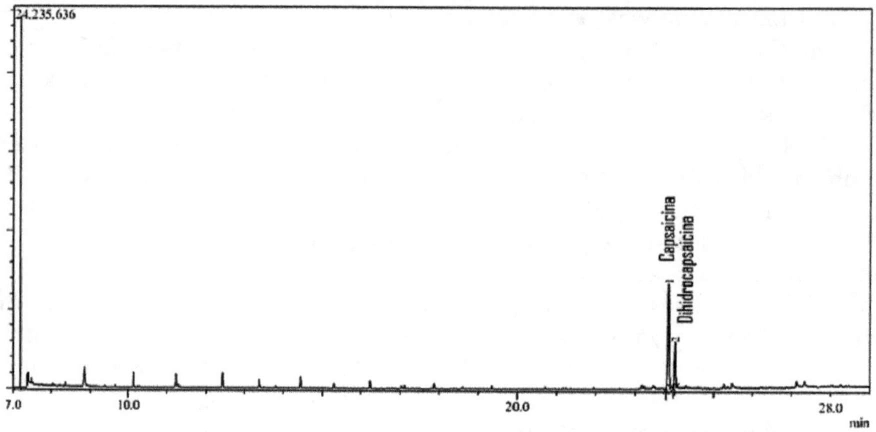

Figure 1.
GC-FID chromatogram of capsaicin and dihydrocapsaicin peaks with retention time around 24 min.

standard solutions made with Sigma-Aldrich Capsaicin (purity capsaicin: 61.1%, dihydrocapsaicin: 31.2% LOT#: LRAA9221 09 September 2015). The regression coefficient R^2 was 0.999 and uncertainty was 0.9%. The subsequent dilution was satisfactory by the statistical linear fit results. In this way, it is possible to quantify the capsaicinoids, that is, capsaicin and dihydrocapsaicin together, on the pepper spray solution.

7. NMR analyses

Nuclear magnetic resonance (NMR) is a powerful analytical tool for identification and quantification of capsaicinoids in raw materials for pepper spray solutions. Its ability to analyze samples with minimum pre-treatment allows a safe analysis that preserves the original characteristics. In the specific case of organic molecules, their sensitivity is increased by the detection of the isotope ^{1}H. Thus, the ability of NMR to evaluate the quality control of the nonlethal arms industry becomes clear.

In order to verify the N.Cap content, a sample of 26.4 mg was weighed and solubilized in 600 µl of CDCl3 in a 5-mm NMR tube. The analysis was performed at 25°C using glass capillary for the quantification of the maleic acid present in the solution with 100 mmol of D_2O. The broad band inverse (BBI) probe inserted into Bruker Avance III 11.75 T equipment was used. The simple pulse sequence used was zg. The spectra were processed with 0.3 Hz line broadening and zero filling.

Figure 2 shows the graphical results, and the relative chemical shift integration of hydrogens of capsaicin and dihydrocapsaicin are presented. All the signals of the main chemical groups present in the sample were identified through the NMR analyses. Groups 1 from capsaicin and 2 from dihydrocapsaicin (amplified in **Figure 2**) can be used for relative integration of the signals. In addition to being unobstructed, they had a higher signal-to-noise ratio due to having six hydrogens contributing to the integration of each signal. Thus, it was possible to verify the proportion of these compounds among themselves present in the mixture. This led to the proportion of 33% dihydrocapsaicin (minority) and 66% capsaicin (majority), the latter being twice the concentration of the former.

The ^{1}H NMR technique proved efficient in the identification and relative quantification of the components present in the mixture of pepper derivatives. The spectra were acquired quickly and with a simple sample preparation step.

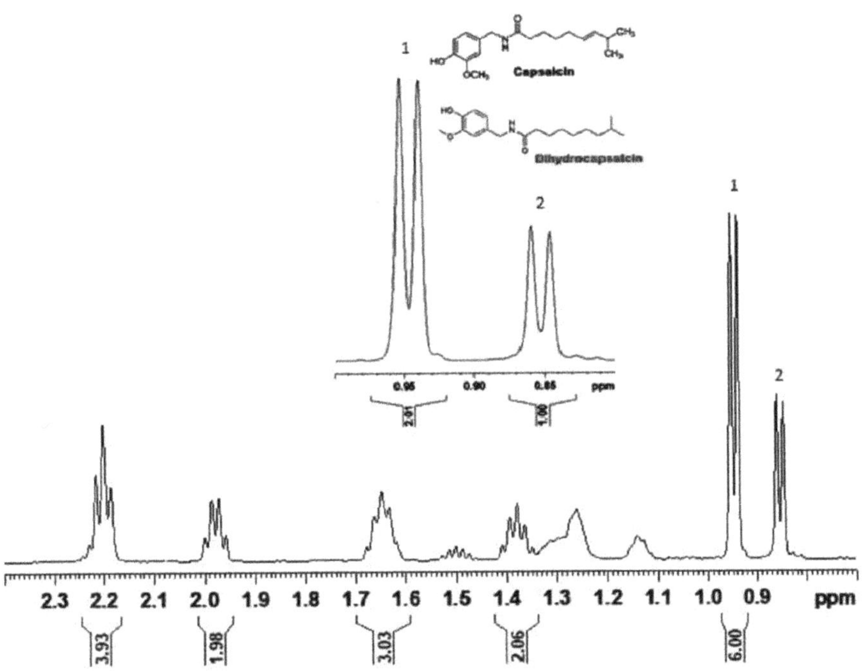

Figure 2.
^{1}H NMR identification and relative quantification of the components present in the N.Cap.

The analyses allowed a broad view of the absence of other compounds, even if minority, thereby proving the purity of the material and absence of degradation products. On the other hand, the presence of other components could be identified concomitantly in a mixture of higher complexity. Thus, this work suggests that ^{1}H NMR can be used as a quality control tool in the nonlethal industry.

8. Conclusion

The main aspects of pepper spraying technology were discussed and the main concepts related to use, hazardous, solution formulations and quantification tests were pointed out. This technology is relatively recent, and a wide field of research is not addressed yet. The advance showed here is around the use of capsaicinoids to extract raw material N.Cap instead of OC to produce pepper spray. The absence of oils and resin is the main point, as nonflammability is achieved much more easily and without the use of hazardous organic solvents. Chemical analyses by GC-MS/FID applied to quantify capsaicinoids (capsaicin and dihydrocapsaicin together) have been demonstrated and can be reproducible for the quality control of this product. The ^{1}H NMR provides information about raw material and how to access quality information and impurity with our relative concentration.

The concentration range between 0.2 and 0.3% (mass of solution) of capsaicinoids guarantees satisfactory spray efficiency, allowing the aggressor to be incapacitated without serious injury. Toxicological study preferred with human voluntaries is the way for a better evaluation of limits.

Author details

Nilton Oliveira Junior[1*], Valter Padulla[1], Vinnicius Ferraço[2] and Giovana Ferronato[2]

1 Condor Non-Lethal Technology S/A, Rio de Janeiro, Brazil

2 Instituto Senai de Inovação em Química Verde, Rio de Janeiro, Brazil

*Address all correspondence to: nilton.junior@condornaoletal.com.br

References

[1] Sutherland RG. Chemical and biochemical non-lethal weapons: Political and technical aspects. SIPRI Policy Paper. 2008;**23**:1-40

[2] NATO. NATO Policy on Non-Lethal Weapons [Online]. 2008. Available: https://www.nato.int/cps/en/natohq/official_texts_27417.htm?selectedLocale=en [Accessed: 19 March 2020]

[3] Office of the United Nations High Commissioner for Human Rights (OHCHR). Basic Principles on the Use of Force and Firearms by Law Enforcement Officials [Online]. 1990. Available: https://www.ohchr. org/en/professionalinterest/pages/useofforceandfirearms.aspx [Accessed: 19 March 2020]

[4] Olajos EJ, Stopford W. Riot Control Agents: Issues in Toxicology, Safety, and Health. USA: CRC Press; 2004. ISBN: 9780415299022

[5] Brown L, Takeuchi D, Challoner K. Corneal abrasions associated with pepper spray exposures. The American Journal of Emergency Medicine. 2000;**18**(3):271-272

[6] Haar RJ, Iacopino V, Ranadive N, Weiser SD, Dandu M. Health impacts of chemical irritants used for crowd control: A systematic review of the injuries and deaths caused by tear gas. BMC Public Health. 2017;**17**:831

[7] Kearney T, Hiatt P, Birdsall E, Smollin C. Pepper spray injury severity: Ten-year case experience of a poison control system. Prehospital Emergency Care. 2014;**18**:1-6

[8] Recer GM, Johnson TB, Gleason AK. An evaluation of the relative potential public health concern for the self-defense spray active ingredients oleoresin capsicum, o-chlorobenzylidene malononitrile, and 2-chloroacetophenone. Regulatory Toxicology and Pharmacology. 2002;**36**(1):1-11

[9] Watson WA, Stremel KR, Westdorp EJ. Oleoresin capsicum (CAP-STUN) toxicity from aerosol exposure. Annals of Pharmacotherapy. 1996;**30**:733-735

[10] Granfield J, Onnen J, Petty CS. Pepper spray and in-custody deaths. RCMP Gazette. 1994;**56**(11):12-17

[11] Broadstock M. What Is the Safety of "Pepper Spray". Vol. 1. Christchurch: NZHTA Tech Brief Series; 2002

[12] Reilly CA, Crouch DJ, Yost GS, Fatah AA. Determination of capsaicin, dihydrocapsaicin, and nonivamide in self-defense weapons by liquid chromatography–mass spectrometry and liquid chromatography–tandem mass spectrometry. Journal of Chromatography A. 2001;**912**:256-267

[13] Lalanne GM, Álvarez AJH, Fernández MJ, Azuara E. Oleoresins from *Capsicum* spp.: Extraction methods and bioactivity. Food and Bioprocess Technology. 2017;**76**:51-76

[14] Junior NO, Padulla V, Flores I, Annunciação D, Neiva M, Brant V. Evaluation of natural capsaicin (N.Cap) in pepper spray by GC-MS/FID, NMR and HPLC as an alternative to the use of oleoresin capsicum (OC). Human Factors and Mechanical Engineering for Defense and Safety. 2019;**3**. Article number 8

[15] Govindarajan VS, Salzer UJ. Capsicum—Production, technology, chemistry, and quality—Part II. Processed products, standards, world production and trade. C R C Critical Reviews in Food Science and Nutrition. 1986;**23**:207-288

[16] Reyes ML. Chemical and pharmacological aspects of capsaicin. Molecules. 2011;**16**:1253-1270

[17] Antonio AS, Wiedemann LSM, Veiga Junior VF. Food pungency: The evolution of methods for capsaicinoid analysis. Food Analytical Methods. 2019;**12**:1327-1345

[18] Govindarajan VS, Salzer UJ. Capsicum—Production, technology, chemistry, and quality—Part II. Processed products, standards, world production and trade. C R C Critical Reviews in Food Science and Nutrition. 1985;**22**:109-176

[19] Zamora AG, Campos ES, Ortega JGL, Morales RP, Ortiz JCR, Hernández JLG. Characterization of different capsicum varieties by evaluation of their capsaicinoids content by high performance liquid chromatography, determination of pungency and effect of high temperature. Molecules. 2013;**18**:13471-13486

[20] Simonsen HT. Short commentary: Capsaicin—The hot and spicy diet turned mild and effective by glycosylation. Biochemistry Insights. 2009;**2**:47-49

[21] Chung M-K, Campbell JN. Review: Use of capsaicin to treat pain - Mechanistic and therapeutic considerations. Pharmaceuticals. 2016;**9**:66

[22] Scoville WL. Note on capsicums. The Journal of the American Pharmaceutical Association. 1912;**1**:453-454

[23] Othman ZAA, Ahmed YBH, Habila MA, Ghafar AA. Determination of capsaicin and dihydrocapsaicin in capsicum fruit samples using high performance liquid chromatography. Molecules. 2011;**16**:8919-8929

[24] García-Casal MN, Peña-Rosas JP, Gómez-Malavé H. Sauces, spices, and condiments: Definitions, potential benefits, consumption patterns, and global markets. Annals of the New York Academy of Sciences. 2016;**1379**(1):3-16

[25] Glinsukon T, Stitmunnaithum V, Toskulkao C, Buranawuti T, Tangkrisanavinot V. Acute toxicity of capsaicin in several animal species. Toxicon. 1980;**18:**215-220

[26] W. J. Jr. Final report on the safety assessment of *Capsicum annuum* extract, *Capsicum annuum* fruit extract, *Capsicum annuum* resin, *Capsicum annuum* fruit powder, *Capsicum frutescens* fruit, *Capsicum frutescens* fruit extract, *Capsicum frutescens* resin, and Capsaicin. International Journal of Toxicology. 2007;**26**:3-10

[27] Saito A, Yamamoto M. Acute oral toxicity of capsaicin in mice and rats. The Journal of Toxicological Sciences. 1996;**21**:195-200

[28] Fuller RW. Pharmacology of inhaled capsaicin in humans. Respiratory Medicine. 1991;**85**:31-34

[29] Chan TC, Vilke GM, Clausen J, Clark RF, Schmidt P, Snowden T, et al. The effect of oleoresin capsicum "pepper" spray inhalation on respiratory function. Journal of Forensic Sciences. 2002;**47**:299-304

10

Capsicum Seeds as a Source of Bioactive Compounds: Biological Properties, Extraction Systems and Industrial Application

Javier Echave, Antía G. Pereira, María Carpena, Miguel Ángel Prieto and Jesús Simal-Gandara

Abstract

Recent research has substantially focused on residual subproducts containing chemical compounds with bioactive properties. Even though there are some culinary or medicinal uses of *Capsicum* seeds, there is still a seed mass waste from pepper processing. Many pepper leading producer countries generally lack the facilities and infrastructure required for such processing technologies and so, pepper seeds are usually either destroyed or employed as landfilling or as animal feed. This involves an inadvertent economic loss for producers as well as a detrimental environmental impact. However, there is a hidden potential within the pepper processing industry related to valorization of pepper seeds to obtain added value by-products and thus reduce generated waste. Pepper seeds are a good source of antioxidants, carotenoids, phenolic acids, flavonoids, and vitamins C, E, and A and are also rich in volatile compounds, among others. The unique alkaloids of this genus are capsaicinoids and capsainoids, which have been linked to many beneficial biochemical and pharmacological effects including anti-oxidative or anti-inflammatory activities. Other prominent bioactive compounds of peppers seeds include saponins, lectins, and polyunsaturated fatty acids. In this context, an overview of the biological properties, extraction systems, and possible industrial application of bioactive compounds of pepper whole fruit and seeds is presented.

Keywords: *Capsicum* spp., seeds, extraction, bioactive, activity

1. Introduction

1.1 Characteristics of the genre

The genus *Capsicum* belongs to the Solanaceae family, which consist of different variants of peppers that can be easily recognized by their size, shape, color and degree of pungency. This latter characteristic allows the classification of chili peppers depending on their Scoville heat units (SHU), a measurement of their pungency, into spicy or hot foods (generally those with small fruits) and sweet or non-pungent (generally those with large fruits). This aspect is very interesting for marketing them and makes them a demanded product [1]. Considering this, we can consider for example the pungency of paprika peppers (10–30 parts per million capsaicinoids), chili peppers (30–600 parts per million) and red peppers (600–13,000 parts per million) [2]. Ripe fruits display a range of colors varying from white to deep red. Likewise, the intensity of red color and the degree of pungency are valued as major quality parameters. That is why the fruit is harvested when its red color has completely developed in order to ensure the highest quality of these features [3].

Capsicum is a very homogeneous genus with 33 species and 10 varieties (**Table 1**), with only five of them being domesticated: *Capsicum baccatum* L., *Capsicum pubescens* Ruiz & Pav., *Capsicum frutescens* L., *Capsicum chinense* Jacq. and *Capsicum. annuum* L. (**Figure 1**) native to Central and South America [5]. Each one of them was farmed independently in Pre-Columbians times in diverse regions of the American tropics [4]. They have been genetically modified to obtain varieties with agronomic interests, as in the case of *C. annuum* L. in which varieties more resistant to adverse factors were created, with a higher content and/or yield of the compounds of interest and fruit quality [6].

Amid the species *C. annuum*, there are different well know varieties like paprika, cayenne, jalapeños or chiltepin. Inside the specie *C. frutescens* are tabasco chilis; among *C. chinense*, the hottest chills (naga, habanero and Scotch bonnet) and among *C. pubescens* and *C. baccatum*, peppers emblematic from South America like rocoto or aji [7].

Although the characteristics of each species differ from each other (**Table 2**), it can be said that the genus *Capsicum* is characterized by being an annual or perennial herb or undershrub with entire or repand leaves that alternate. Its flowers are pedicelled, enclosing five petals; it is axillary and can appear solitary or in groups of two or three. Its sepals connate in a subentire or minutely five-toothed calyx that is much shorter than the fruit. It also has 5 stamens, which are adnate nearly to base of corolla-tube and are characterized for being short filaments. Its carpels connate in a two-celled (in some cases three) ovary, and its anthers are dehiscence longitudinal

Kingdom	**Plantae**
Subkingdom	Tracheobionta
Division	Magnoliophyta
Class	Magnoliopsida
Subclass	Asteridae
Order	Solanales
Family	Solanaceae
Genus	*Capsicum*
Species	*C. baccatum*, *C. pubescens*, *C. frutescens*, *C. chinense*, *C. annuum*
Varieties	*C. annuum* var. *annuum*, *C. annuum* var. *glabriusculum*, *C. baccatum* var. *pendulum*

Table 1.
Taxonomic classification of the genus Capsicum *[1, 4].*

Figure 1.
Some of the major cultivated pepper species. (A) Capsicum frutescens*; (B)* Capsicum annuum*; (C)* Capsicum chinense*, and (D)* Capsicum baccatum.

Species	Flower color	N° flower	Seed color	Calyx constriction	Distribution
C. annuum	White	1	Tan	Absent	Colombia, USA
C. frutescens	Green	2–5	Tan	Absent	No studies
C. chinese	White-green	2–5	Tan	Present	South America
C. baccatum	White with yellow spot	1–2	Tan	Absent	South America
C. pubescens	Purple	1–2	Black	Absent	South America

Table 2.
Characteristics that allow to distinguish between the most common species of the Capsicum *genus [8].*

without any exceeding filaments. As for fruit and seeds, the first is irregularly shaped (globose or elongate) with many seeded berries. Seeds are discoid, smooth or subscabrous [8].

Thus, its physical appearance defers from one species to another. For example, *Capsicum annuum* is an annual cultivate that reaches a height of 1 m and has glabrous or pubescent lanceolate leaves, white flowers, and fruit of varying length, color, and pungency depending upon the cultivar and growth conditions. It also the most widely cultivated pepper species around the world. Considering another species, *Capsicum frutescens* is a short-lived perennial with woody stems that reach a height of 2 m, with glabrous or pubescent leaves, with two or more greenish-white flowers per node, and extremely pungent fruit [2].

Being a cold sensitive plant, the best conditions for production are between 7 and 29°C and an annual precipitation of 0.3–4.6 mm. It grows best in well-drained, sandy or silt-loam soil and a soil pH of 4.3–8.7. Hot and dry weather is also desirable for fruit ripening [2]. To carry out its cultivation it is necessary to seed or transplant the peppers, harvesting 3 months after planting [2].

1.2 *Capsicum* seeds sources and production

Chili (a variety of *C. annuum*) is one of the first plants cultivated in Mesoamerica, existing evidence of its use for the last 9000 years. It can be used in multiple ways: fresh, dry, as a spice powder, natural dye, antioxidant, bactericide and fungicide, as a drug in the pharmaceutical industry, in the cosmetology industry or in food industry (sausage, canned meats) [9]. These characteristics made them an essential part of daily cooking in many Latin American and Asian countries, for example, curry blends in India or in many meat sausages both fresh and dehydrated in the Mediterranean region [3].

On the basis of extracts obtained from pre-ceramic in the Coxcatlan caves, it is believed that the domestication of *C. annuum* probably occurred in the northeast or central-east of Mexico, being older than the remains of corn, beans and pumpkin [1]. In fact, peppers presently represent one of the vegetables of greater economic importance just behind the tomato for several American tropical countries [10].

Of the five domesticated species of *Capsicum* spp., the largest cultivated and with higher production per hectare are *C. annuum* and *C. frutescens* [11]. As for its world production, it is estimated that it is nearby 24.9 million tons (*C. annuum*), making it the ninth most produced vegetable in the world. Likewise, it has had an annual average growth of 6.26% in the last 10 years, with Mexico being its main consumer with an annual average of 8 kg/person. Its main producers are China (60.6%), Turkey (8.4%), Mexico (7.8%), Spain (5.0%), USA (4.3%), Indonesia (3.4%), Nigeria (2.8%), Egypt (2.2%), Korea (2.0%) and Italy (1.7%) [7]. In Mexico, one of the countries with the highest production levels, there are about a hundred cultivars of hot *Capsicum* spp. with varying degrees of spiciness, size, shape and colors. Nonetheless, there are also varieties that are cultivated throughout the world. This is the case of *C. annuum* or *C. chinense* [12]. Hence, it is a product that is traded internationally. This means that a series of elements such as aflatoxins, pesticides, residuals, microbial contaminations or infections, capsaicin levels and color values are subjected to inspection. All these aspects are controlled by law in several countries (EU, USA) [8].

Pepper quality depends on their composition, which is determined by factors such as environmental cultivation conditions, variety, ripeness, and pre-harvest and post-harvest handling and preservation [13]. The degree of ripening required may be one of the most important factors in quality, but it will also depend on the destined market, since not the same degree of maturity is desired for all the possible uses. Notably, the moment of harvesting is also important for the maintenance of the quality as metabolic activity persists after harvesting [14].

1.3 Chemical composition

Regardless of the enormous consumption and production of this kind of vegetables, there is little data about the chemical composition of the different varieties (**Table 3**). However, the demand and cultivation of peppers, especially "hot" cultivars, has increased due to its flavoring and medicinal properties. Some of the latter have been described as anticancer, antioxidant and antimicrobial. Its edible and

Common name	State	Carbohydrates	Protein	Fat	Capsaicinoids	Fiber	Ash
Guajillo	D	58.00	12.89	12.43	5.97	Nd	7.52
Ancho	D	60.21	12.05	9.82	8.50	Nd	7.81
Pasado	D	66.18	12.61	5.41	9.74	Nd	7.18
Pasilla	D	60.53	12.28	13.76	11.80	Nd	5.85
Puya	D	63.76	13.25	8.11	12.13	Nd	7.82
M. Tres venas	D	61.05	13.28	9.61	14.40	Nd	7.02
Chiplote Meco	D	57.68	15.22	9.08	29.01	Nd	9.54
Jalapeno	D	63.97	14.36	4.23	58.40	Nd	7.32
Mirasol	D	58.96	14.05	7.49	58.55	Nd	9.61
Morita	D	58.91	14.12	7.60	67.32	Nd	8.59
Serrano	D	67.93	12.78	2.26	102.73	Nd	5.81
Chiplote	D	62.92	12.72	8.66	143.57	Nd	6.92
De Arbol	D	59.41	12.75	13.38	193.51	Nd	8.82
Piquin	D	62.25	13.72	11.02	368.83	Nd	7.28
Habanero	D	61.13	13.52	4.63	1312.10	Nd	7.51
Marako fana	D	35.3	11.8	11.2	Nd	27.3	5.3
Bako local	D	39.5	8.7	9.5	Nd	26.0	7.3
Oda haro	D	37.1	9.2	9.2	Nd	28.6	7.3
Arnoia red	F	6.23	0.15	0.54	Nd	1.62	0.62
Arnoia green	F	3.84	0.14	0.22	Nd	1.63	0.40
Arnoia green B	F	3.51	0.12	0.16	Nd	1.31	0.33
Hot pepper	D	Nd	21.29	23.65	Nd	38.76	4.94
Chunhamuchuk	D	Nd	15.05	29.27	Nd	48.72	3.59
Amhanegosa	D	Nd	14.66	26.70	Nd	52.10	3.28
Hanbando	D	Nd	14.08	27.84	Nd	38.43	3.49
Dachon I	D	Nd	15.99	19.53	Nd	50.61	3.76
Samgang	D	Nd	13.90	23.50	Nd	50.71	3.71
Chunhajeil	D	Nd	14.67	21.87	Nd	52.54	3.46
Daejangbu	D	Nd	14.88	26.50	Nd	53.78	3.47
Hongjangkun	D	Nd	15.17	21.61	Nd	54.66	3.33
Kumbit	D	Nd	15.09	23.28	Nd	46.17	3.18
Dokyachungjung	D	Nd	15.75	25.13	Nd	53.36	3.86
Dangchan	D	Nd	15.36	19.99	Nd	55.63	3.46
Chohyang	D	Nd	13.25	18.05	Nd	59.13	3.77
Taesan	D	Nd	14.71	23.45	Nd	48.80	3.11
Ganggun	D	Nd	15.55	20.63	Nd	52.71	3.43
Chungsan	D	Nd	15.06	20.45	Nd	50.75	3.05
Dachon II	D	Nd	15.89	18.83	Nd	45.73	3.61
Wangdaebak	D	Nd	16.53	23.65	Nd	54.39	3.28
Chunhailpum	D	Nd	15.70	19.79	Nd	53.34	3.38

Common name	State	Carbohydrates	Protein	Fat	Capsaicinoids	Fiber	Ash
Daechan	D	Nd	13.88	20.64	Nd	54.68	3.72
Mixed	D	Nd	14.01	24.09	Nd	50.26	3.26
Sandia	D	Nd	14.95	23.07	Nd	58.34	3.22
R-Naky	D	Nd	14.36	23.57	Nd	60.19	3.57
New Mexico 6	D	Nd	14.79	21.95	Nd	60.61	3.48
LB-25	D	Nd	14.87	25.06	Nd	52.98	3.29

F, fresh; D, dried; Nd, not described.

Table 3.
Different Capsicum *spp. and their proximate chemical composition (g/100 g) [14–18].*

nutritional value is acknowledge as well, since it is rich in vitamins (A, C, B_6, E), carotenoids (β-carotene), flavonoids, oils, oleoresins and alkaloids [19]. Therefore, the compounds that can be found in this genre are carbohydrates (accounting for approximately 85% of dry weight), polyphenols (0.5% of dry weight) and important molecules such as capsaicinoids, carotenoids and vitamins [20]. Given these facts, peppers are considered a good source of most essential nutrients [14].

The capsaicinoids content, which depends on the variety and maturation stage, will determinate the pungency. *C. chinense* and *Capsicum annuum* var. *aviculare* contain larger amounts of capsaicin and dihydrocapsaicin (ratio of 2:1), while some varieties of *Capsicum annuum* var. *annuum* showed an average proportion of 1:1. These variations could be attributed to environment, genetics and extraction methodologies [15]. Even though in some cases peppers are sought to be spicy, this can also be a limitation. This is the case, for example, in obtaining dyes from this raw material in the food industry. In this case, extraction methods need to be improved to prepare non-pungent oleoresins from pungent *Capsicum* fruits. This is achieved by the selective removal of capsaicinoids which allows the exploration of a large number of pungent varieties with good oleoresin yielding [20].

Capsaicinoids are the characteristic pungent compounds of the *Capsicum* genus. They include capsaicin, dihydrocapsaicin, nordihydrocapsaicin, homocapsaicin, and homodihydrocapsaicin. Capsaicin is the most abundant. As an example, capsa-icin constitutes together with dihydrocapsaicin approximately 90% of total capsaicinoids content of chili peppers [15].

The characteristic red color of many peppers is determined by the presence of different carotenoids. Actually, more than 50 different carotenoids can be found in this kind of material. Some of them are capsanthin, capsorubin, and cryptocapsin which give brilliant red color (ripe fruits) or β-carotene, violaxanthin, zeaxanthin and β-cryptoxanthin which give yellow–orange color [3]. Nevertheless, the color will depend on the state of maturity. For example, jalapeño (*C. annuum*) has a green color, and when it is mature it presents an intense red color [7]. Therefore, depending on the degree of maturity, the physicochemical parameters, flavor and mineral composition differs. For example, fat, protein, ascorbic acid (vitamin C), soluble solid content and titratable acidity will increase during ripening [14].

Using jalapeño as an example to study composition, its main component is water, then carbohydrates (5.3%), fiber (2.3%), protein (1.2%), fat (0.1%) and minerals, being the most important potassium (340 mg per 100 g of fresh product). It also has calcium (25 mg per 100 g of fresh product), magnesium (25 mg per 100 g of fresh product), sodium (7 mg per 100 g of fresh product), iron (2 mg per 100 g of fresh product) and zinc (0.3 mg per 100 g of fresh product). As for vitamins, the most important are ascorbic acid, retinol and folic acid (72, 20 and 23 mg per 100 g of fresh product, respectively). Other vitamins that can be found in peppers are thiamine, riboflavin, niacin and pyridoxine. Moreover, jalapeños also contain important amino acids such as lysine, methionine and valine (252, 40 and 23 mg per 100 g of protein, respectively) [7].

2. *Capsicum* seeds as a source of bioactive compounds

Like many fruits and plants, peppers are an excellent nutritional source. As aforementioned, it has an abundance of minerals, vitamins, aminoacids, carotenoids as also phytochemicals like phenolic compounds or polyunsaturated fatty acids (PUFAs). Likewise, capsaicins are unique to the genus *Capsicum*, being responsible of the pungency of many pepper species. Capsaicinoids have demonstrated to induce a vast range of bioactivities, such as anti-inflammatory, anticancer analgesic, antimicrobial, hypotensive and induce lower adipogenesis or lower body temperature by isolated action or in synergy with other compounds [21–25].

Different bioactive compounds have been isolated and extracted from *Capsicum* spp. fruits and seeds and its concentrations vary among species. More information on reported compounds studied may be found in (**Table 4**).

2.1 Phenolic compounds

Phenolic compounds, also referred as phenolics, are secondary metabolites that may be found in a wide spectrum of plant species. They are synthesized as a result of adaptation to biotic and abiotic stress through the phenylpropanoid pathway playing an important role in plant development, because they act as a defensive mechanism that eases plant growth against harsh conditions [33, 34]. Based on recorded knowledge, phenolic compounds exhibit numerous potential health benefits that are already well described in scientific literature and are currently a major current focus of nutritional and pharmacological research [35, 36]. Phenolic acids and flavonoids are the main phenolic phytochemicals found in peppers. Likewise, capsaicinoids are synthetized in the same biochemical pathway and exhibit some similar properties such as antioxidative activity [34].

The yield of phenolic compounds recovered from an extraction can be very different, depending largely on the extraction method, the conservation of vegetal material and maturity state [37]. Generally, fresh raw material preserves the highest quantities of phenolic compounds [16]. Several factors contribute to yielding disparity, such as heterogeneous genotypes, growing and harvesting conditions of the samples [38]. Some of the most prominent non-capsaicinoid phenolic compounds because of their valuable health benefits are phenolic acids like gallic acid, caffeic acid, chlorogenic acid or ellagic acid (**Figure 2**). Flavones are another important group of phenolic compounds, being some of the most prominent kaempferol, quercetin, luteolin or rutin (**Figure 3**) [26]. These phenolics have demonstrated great health benefits and many of them are commercially available in purified products extracted from other plant species [39, 40]. Even so, their concentration differs among species but not much among varieties [26]. Other phenolic compounds are reviewed in the next paragraphs as in the case of some carotenoids or vitamins.

Regarding capsaicinois, they are synthesized naturally in the placenta of pepper fruits by enzymatic transformation of vanillylamine, the phenolic portion of the molecule, which confers this alkaloid its antioxidant capacity [41]. The seeds are not the primary source of capsaicinoids but they may absorb them because they are in close proximity to the placenta, which is the richest capsaicin fraction [42]. Their presence in the seed and the high concentrations they achieve, has been observed to

Species and varieties	Bioactive compounds						
	Major phenolic compounds	**Vitamins**	**Pigments**	**Volatiles**	**PUFAs**	**Peptides**	**Ref.**
C. annuum	4-Hydroxybenzoic acid, vanillic acid, caffeic acid, *p*-coumaric acid, sinapic acid, ferulic acid, ellagic acid, naringein	Nd	Nd	Nd	Nd	Nd	[26]
C. annuum	Elagic acid, gallic acid, chlorogenic acid, caffeic acid, salicylic acid, rutin	Nd	Nd	Nd	Nd	Nd	[21]
C. annuum	Caffeic acid, gallic acid, ferulic acid, rutin, capsaicin, dihydrocapsaicin	Tocopherols	Capsanthin, zeaxanthin, capsorubin, lutein, β-carotene, cryptocapsin	Nd	Linoleic, palmitic, oleic acids	Nd	[27]
C. annuum	Nd	Nd	Nd	Nd	Nd	Mannose/ glucose Specific lectin	[28]
C. annuum	Nd	Ascorbic acid	Nd	Betulin, β-sitosterol	Oleic, palmitic, linoleic, elaidic acids	Nd	[29]
C. annuum	Gallic acid, caffeic acid, ferulic acid, chlorogenic acid, cathechin, epicathechin capsaicin, dihydrocapsaicin	Ascorbic acid	Nd	Nd	Nd	Nd	[30]
C. annuum var. *annuum*	Caffeic acid, cinnamic acid, coumaric acid, capsaicin, vanillin	Nd	Nd	Nd	Nd	Phenyl-alanine	[31]
C. baccatum	Tannins, flavonoids, coumarins, polyphenols	Nd	Nd	Saponins	Nd	Amino acids	[32]

Nd, not described.

Table 4.
Capsicum *seeds bioactive compounds.*

rise the riper the pepper is [43]. It is also confirmed that fresh seeds yield more capsaicin than dry seeds, which suggests that surface capsaicinoids are sensitive to heat and/or oxidation [29]. "Hot" pepper cultivars attribute their pungency to high levels of capsaicinoids whereas non-pungent or "sweet" peppers (e.g., bell pepper) have very low capsaicinoids quantities [44].

Capsinoids are other non-pungent capsaicin analogues noticeably found in *C. baccatum* var. *praetermissum* and the sweet cultivar CH-19 of *C. annuum* [45, 46].

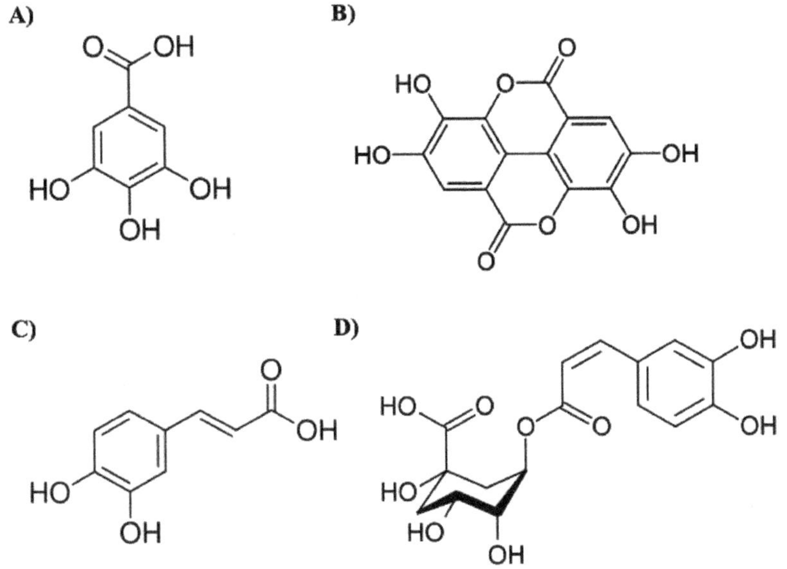

Figure 2.
Chemical structure of some main phenolic acids found in peppers: (A) gallic acid; (B) ellagic acid; (C) caffeic acid and (D) chlorogenic acid.

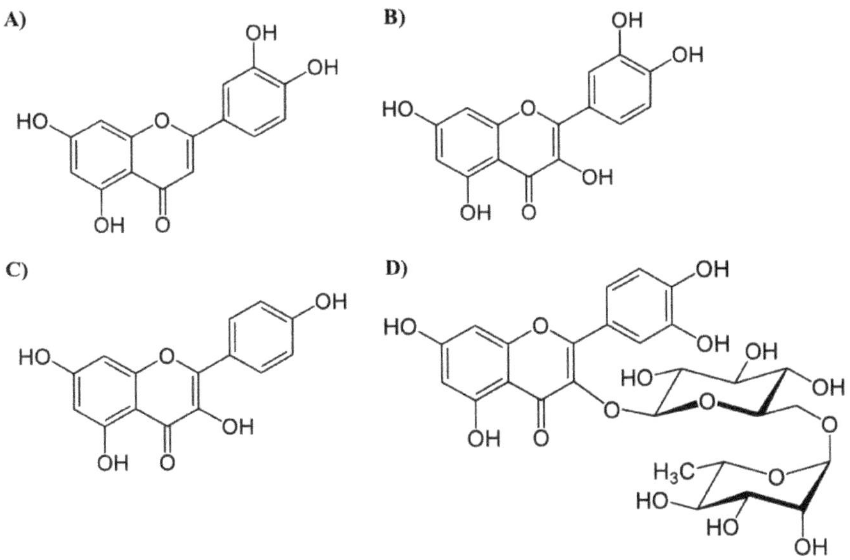

Figure 3.
Chemical structure of the main flavonoids found in pepper: (A) luteolin; (B) quercetin; (C) kaempferol; and (D) rutin.

Capsinoids exhibit biological activities similar to capsaicin due to being capsaicinoids analogs, except that they display much lower pungency in comparison [45]. Capsaicinoids and capsinoids manifest many promising therapeutic properties such as apoptosis induction, antioxidation, anticancer (cytostatic), or immunomodulation, making them attractive targets for pharmaceutical research [24, 47].

Besides, it is worth mentioning that the well-known flavone chrysoeriol (**Figure 4**), present in a multitude of vegetables and at least in *C. annuum* and *C. frutescens* fruit and seeds, has been determined to inhibit the growth of several microorganisms at very low quantities [31].

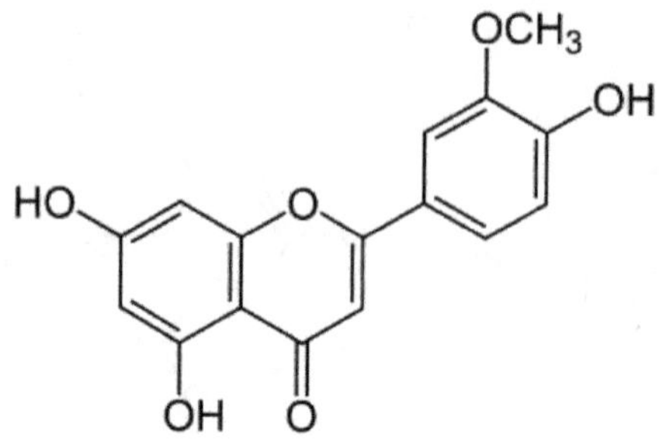

Figure 4.
Chemical structure of chrysoeriol.

2.2 Fatty acids

Peppers are fruits rich in polyunsaturated fatty acids (PUFAs), and their seeds show even greater concentration per gram [42]. The main fatty acids are indeed PUFAs [48]. These PUFAs, which are in the whole fruit and seeds, are mainly linoleic acid, palmitic acid α-linolenic acid and stearic acid [16, 48]. Moreover, peppers appear to have very low levels of saturated fatty acids. Linoleic acid has the highest concentration (≈70%) and the other fatty acids show much lower levels [48, 49]. PUFAs and specifically linoleic acid and α-linolenic acid, are recognized as essential fatty acids and are precursors of other important fatty acids in metabolism like arachidonic acid and eicosapentanoic acid or prostaglandins that in whole contribute to normal physiological performance [50].

2.3 Pigments

Chlorophylls and carotenoids constitute another group of valuable pepper nutrients that affect its color, but they also have important antioxidative, anti-inflammatory effects and promote immune response [22, 51, 52]. Zeaxanthin, β-carotene, violaxanthin, lutein and β-cryptoxanthin are the pigments with highest concentration among *Capsicum* spp. [37, 43]. Capsanthin is another red colored pigment mostly present in red bell pepper (*C. annum*) [53]. They all show strong antioxidative properties and increase their levels the more mature the pepper is [44].

β-carotene and β-cryptoxanthin for instance, possess the added value of being able to be converted to vitamin A [54]. Vitamin A plays a vital role in disease prevention and development [54].

2.4 Vitamins

As it has been mentioned, peppers contain several vitamins like ascorbic acid and tocopherols like α-tocopherol, γ-tocopherol and δ-tocopherol, are all isomers of vitamin E [55, 56]. These essential vitamins contribute to the normal metabolism, promote immune response and also have antioxidative bioactivity [52, 57]. Because of this, ascorbic acid is an essential vitamin that is also used as a natural food preserver [52]. Indeed, peppers contain levels of vitamin C corresponding with those found in many citrus fruits and other vegetables considered good sources of this vitamin [14]. Tocopherols are well-known lipophilic antioxidants and appear to exert a vital part in the normal T lymphocyte maturation, lower age-related increase in tissue inflammation and lower interleukin production [55, 58]. This makes vitamin E an important modulator to an orderly and better inflammatory reaction, among other health benefits [58].

2.5 Volatile compounds

Volatile compounds in *Capsicum* spp. have been studied because of the importance they have in many plants as pollinator attractors and their influence in odor and flavor, which are key features of any edible vegetable [59, 60]. Some of the most valuable are hexyl isopentanoate, hexyl isobutanoate, and β-ionone [61]. Extracts of *C. annuum* have also showed high levels of 4-methoxyphenol, ethylhexadecanoate, hexanal or isopulegol [62].

2.6 Other minor compounds

There are a few other minor and less studied compounds amid the different pepper species, because research has focused almost entirely on finding and studying phytochemicals of metabolic importance. In regard to these bioactive compounds, *C. annuum* or *C. frutescens* saponins have showed antimicrobial activity against several opportunistic and pathogenic fungi genres [63]. Saponins are triterpene glycosides with the ability to disrupt the cell membrane and wall of fungal cells during cell proliferation, which leads to cell lysis [27] As such, saponins do not have effect on non-germinated fungi, but they inhibit its growth. Although there are several present in pepper tissue, CAY-1 has been well-studied (**Figure 5**), presenting a high antifungal activity [60].

Another of these minor chemicals are lectins, a group of proteins with glycoside agglutination properties. Some lectins have been described in *C. frutescens* and *C. annuum* seeds and have been studied as possible antimicrobials and have yielded positive results against some fungi [28, 64].

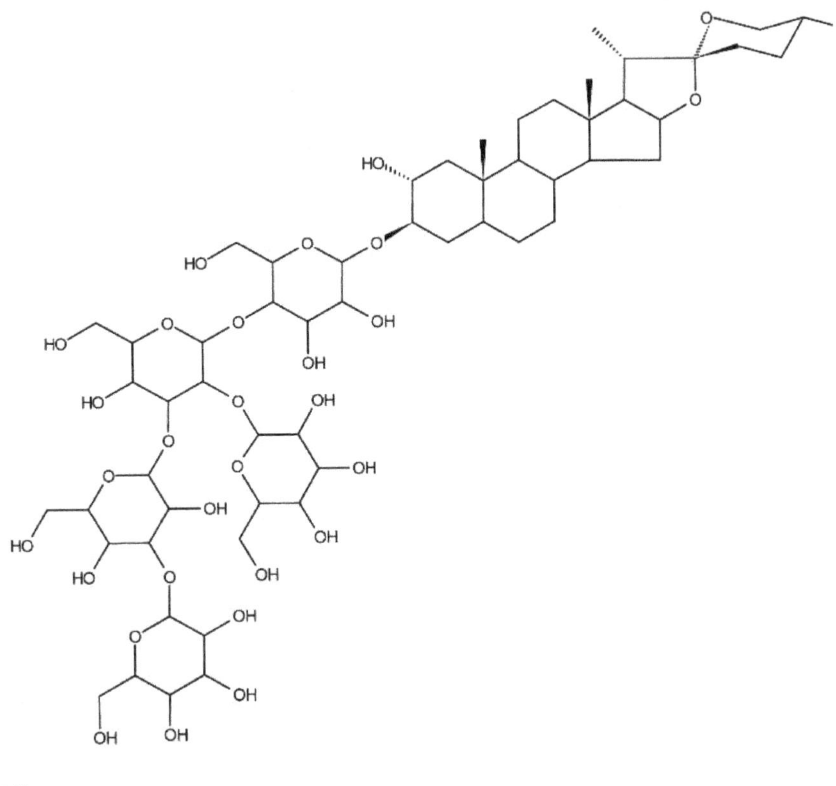

Figure 5.
Chemical structure of CAY-1. Adapted from [60].

3. Bioactivities of the raw material

In recent years, nutraceutical and therapeutic research has focused its view towards both exotic and domestic fruits and vegetables as a source of phytochemicals with the ability to induce beneficial bioactivities [65]. These natural chemicals have been and are already used as a main source of therapeutics in traditional and modern medicine, reaching one third of the total production of therapeutics [66]. Furthermore, processing waste by-products of fruits and vegetables are being researched as a viable source of phytochemicals that would be affordable and reduce economic and ecological impacts of wasted by-products or taking produce out of the food market [67, 68]. Thus, fractions that are not employed in nourishment such as the placenta, seeds or leaves of many species can prove to be a valuable resource instead of end as waste or fertilizer, which is the most common use for vegetable non-edible parts [61].

Taking into account the aforementioned compounds present in the many different species of the genus *Capsicum* and different regional cultivars, as in the case of *C. annuum*, many biological activities are expected to be found in direct or indirect physiological response to the presence of these substances. It is widely known that capasaicinoids, capsinoids, saponins, many phenolic compounds and its valuable nutrients are behind the beneficial effects of this vegetable, which has been traditionally employed as a source of medicinal remedies all across South America, Asia and Africa [63]. The intensity and type of activity may vary depending on the concentration of these compounds and will differ in each species and cultivar of *Capsicum*, the bioavailability of each compound, the ripening state and from which fraction of the vegetable the extracts are obtained [43, 44].

These bioactivities have been described in scientific reports through tissue culture and both animal and human test research as antioxidant, antimicrobial, anti-inflammatory, anticancer, analgesic or even antidiabetic [21, 22, 34]. Considering this with the fact that *Capsicum* spp. is widely grown around the world, it makes it to be considered an excellent source of new therapeutics and dietary supplements that could lead to a healthier well-being. Indeed, many extracts of pepper fruits and seeds with a varying purification degree are marketed and can be purchased [69].

Some of the most recognized bioactivities found by chemicals in the whole pepper and its seeds will be reviewed and a brief compendium of the bioactivity research can be found in **Table 5**.

3.1 Antioxidant

Oxidative stress is caused as a result of the presence of reactive oxygen species (ROS) which may be produced in oxidative metabolism and exposure to the environment [33]. The term ROS englobes the molecules superoxide radical (O_2^{2-}), the hydroxyl radical (OH^-) and hydrogen peroxide (H_2O_2). They are produced by the sequential reduction of molecular oxygen in various metabolic reactions [76]. O_2^{2-} is the most unstable form but it may dismutate to H_2O_2 by the action of endogenous superoxide dismutase (SOD) enzyme reaction or non-enzymatically [76]. The effect of oxidative stress mainly translates into changes in the rate of metabolic reduction reactions and an increased rate of DNA mutations and cell mitosis. They are also main signals indicating cell death, which in turn triggers inflammation via the pro-inflammatory factor nitric oxide (NO), which is released by macrophages [33, 76]. Since these are both cause and result of cancer development, it is consid-ered that high levels of ROS are detrimental to health [40, 65, 76]. All living aerobic

Activity	Bioactives	Species	Type of study	Test results	Ref.
Antioxidant	Capsaicin, dihydrocapsaicin	*C. annuum*, *C. baccatum* var. *pendulum*, *C. chinense*, *C. frutescens*	In vitro	DPPH, ABTS	[68]
	Total phenolic compounds	*C. annuum*	In vitro	DPPH, ABTS, ORAC	[26]
	Phenolic extracts	*C. annuum*	In vitro	DPPH, ferrus chelating	[39]
	Seed oil	*C. annuum*	In vitro	DPPH, ABTS	[42]
Antimicrobial	Capsicosides A, G, D	*C. annuum* var. *acunimatum*	In vitro	Gia. various yeasts	[32]
	Capsaicin	*C. annuum*, *C. baccatum*, *C. chinense*, *C. frutescens*	In vitro	Gia. *Clostridium tetani* & *C. sporogenes*	[31]
	Phenolic extracts	*C. annuum*	In vitro	Gia. *S. aureus*, *L. monocytogenes*, *S. typhimurium*, *B. cereus*	[70]
	CAY-1	*C. frutescens*	In vitro	Gia. *A. flavus*, *A. fumigatus*, *A. parasiticus*, *A. niger*, *C. albicans*, *P. carinii*	[60]
	Lectins	*C. annum*	In vitro	Gia. *A. flavus*, *F. graminearum*	[28]
	Total polyphenol content, capsaicins	*C. annuum*	In vitro	Gia. *E. faecalis*, *P. aeruginosa*, *E. coli*, *S. aureus*	[21]
	Jalapeño extracts	*C. annuum* var. *annuum*	In vitro	Gia. *L. monocytogenes*	[71]
	Capsaicinoids, chrysoeriol	*C. frutescens*	In vitro	Gia *E. faecalis*, *B. subtillis*, *S. aureus*, *P. aeruginosa*, *E. coli*, *C. albicans*	[31]
	Residual aqueous extract	*C. baccatum*	In vitro	Gia *S. epidermidis*, *P. aeruginosa* biofilm	[72]
Anti-inflammatory	Capsaicin	Nd	In vitro	Inhibition of inflammatory transcription factor NF-κB and AP-1	[73]
	Capsaicin	Nd	In vitro	Inhibition of Ikα and Ikβ via NF-κB	[74]
	Capsaicin	Nd	In vitro	Inhibition of adipose tissue inflammation (interleukin 8, c-Jun)	[75]

Nd, not described; Gia., growth inhibition against; DPPH, 2,2-diphenyl-1picrylhydrazyl scavenging method; ABTS, 2,2′-azino-bis(3-ethylbenzothiazoline-6-sulfonic acid) diammonium salt scavenging methods; ORAC, oxygen radical absorbance capacity.

Table 5.
Properties and mechanisms of bioactive compounds and Capsicum *spp. extracts evaluation.*

organisms have developed defense mechanisms against oxidative stress through the synthesis of reductive biochemicals or enzymes such as SOD. The most important antioxidative biochemicals are phenolic compounds, which are prominently found in plants [52, 66]. These antioxidants present in peppers, mainly identified as the mentioned phenolic compounds, vitamins and pigments, have the potential to reduce biological oxidative stress and thus preventing the incidence of many related diseases, but also to further food preservation by inhibiting the oxidative metabolism of decomposing fungi and bacteria [33, 35, 38]. Phenolic compounds are the principal antioxidants in nature and generally show the greatest antioxidative capacities [65]. However, high concentrations of few phenolic compounds showed pro-oxidant effects due to a concentration imbalance between ROS and the phenols, which is why it is important to maintain intake of different phenolic compounds in order to benefit from their antioxidant properties [52]. Thus, antioxidants have a wide extent of applications, and many natural and synthetic antioxidants have been used by the food industry in order to better preserve raw or precooked products that would otherwise have a much shorter shelf-life [22].

The antioxidative effect of the aforementioned phenolic compounds essentially works by scavenging free superoxide and hydroxyl radicals and thus preventing high levels of ROS, NO and oxidation of sensitive biomolecules like proteins or lipids (**Figure 6**) [36, 40]. This results in a better physiological performance, immunomodulation and DNA mutation protection. Furthermore, the decrease of oxidative stress results in better cardiovascular health. This is due to the fact that low-density lipoprotein (LDL), as the main cholesterol carrier in the circulatory system, is susceptible to oxidation by ROS [58]. Oxidized LDL presence has been found to be cause of atherosclerosis, a vascular ailment caused by the formation of plaques inside the arteries that may weaken blood flow and lead to cardiovascular diseases [77]. On top of that, pepper and its seeds are described to be one of the richest vegetables in phenolic compounds [29]. Then, it should be taken into account the antioxidant properties of the many phenolic acids, capsaicinoids, vitamins and pigments present among the *Capsicum* genus and the fact that these exploitable compounds are available not only in the fruit, but also in the non-edible fractions [39, 57, 78].

3.2 Antimicrobial

Pepper seed extracts and selected pigments, phenols, capsaicins and capsiates have been tested against the most common microorganisms present in foods and/or potential pathogens. The major antimicrobial effect of capsaicinoids has been found to be against common opportunistic and pathogenic fungi like *Fusarium* spp., *Aspergillus* spp. and yeasts like *Saccharomyces cerevisiae* or *Candida albicans*, but also against certain bacteria like *Streptococcus* spp., *Helycobacter pylori*, *Pseudomonas aeruginosa* or

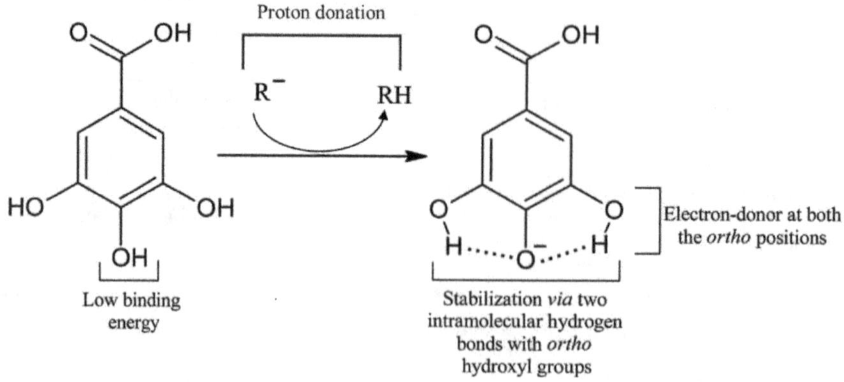

Figure 6.
Scheme of radical scavenging performed by gallic acid as an example of phenolic compounds. Adapted from [36].

Listeria monocytogenes [21, 32, 38, 79]. It has been reported that it can also reduce the virulence of pathogenic infections by *Vibrio cholerae* or *L. monocytogenes*, by inhibiting pro-inflammatory factors or toxin production [79–81]. There are few *in vivo* studies of these effects, though there are some reports of poultry showing less hepatic damage from *Salmonella enteritidis* when given feed supplemented with capsaicinoids, which could be related to its antioxidative activity [24].

Capsaicinoids appear to inhibit, fungi, and both Gram positive and negative bacterial growth in several studies. In all of them, capsaicin and/or dihydrocapsaicin showed to stop or slow colony development in a significant degree [24, 35]. Some interesting results of antimicrobial tests with capsaicinoids reported that it inhibits the growth of *Escherichia coli*, *L. monocytogenes*, *Bacillus subtillis* and many strains of *Staphylococcus aureus* [21].

Even though capsaicinoids possess these interesting antimicrobial properties, the positive correlation between antimicrobial activities, concentration and its pungency and potential irritation effects make them less suitable when it comes to applying this compound as a food preserver [46]. Thus, it is interesting to study the effects on this area of alternative chemicals found in pepper fruit and seeds. CAY-1, chrysoeriol and lectins have been previously mentioned as such.

CAY-1 is found in *C. frutescens* and a study demonstrated to produce cell membrane lysis in fungi like *Aspergillus flavus*, *Aspergillus niger*, *Fusarium solani*, *Fusarium moniliforme*, *Pneumocystis carinii* or *Cansida albicans* [60]. It is ineffective though against bacteria though, probably because of the different type of cell membrane and wall [60]. It has also been suggested to combine this saponin with certain metallic elements such as silver in order to produce antimicrobial surfactants [82]. Other saponins isolated from *C. annuum* seeds, like capsicosides, which are furostanol saponins, have showed similar effects on growth inhibition against yeasts [27].

Chrysoeriol is another potent antimicrobial compound present in a variety of plants as well as in peppers [83]. Several studies have presented that chrysoeriol is capable of greatly inhibit the growth of *Enterococcus faecalis*, *Klebsiella pneumoniae*, *S. aureus*, *L. monocytogenes*, *B. subtillis* or *C. albicans* [31, 84]. Given the very low concentrations at which it exerts its antimicrobial effect, it may be an optimal food preserver against many opportunistic and nosocomial microbes.

Certain lectins isolated from peppers have showed antifungal properties, inhibiting growth of some common opportunistic fungi like *Aspergillus flavus* and *Fusarium moniliforme* by limiting surface adherence. However, its inhibitory action is not as effective across all species of these genres [28, 85].

Furthermore, a recent study analyzed the antiadhesive capacity of natural peptides from *C. baccatum* var. *pendulum* on biofilms made by *Staphylococcus epidermidis* [64]. Considering that biofilms are a source of significant virulence, new studies on this field present extremely interesting. Altogether, these findings may contribute to develop new uses for natural antimicrobials, which would be an application of vital priority, given the global concern on the appearance of newborn antibiotic resistant microorganisms [31, 84].

3.3 Anti-inflammatory

Although an anti-inflammatory action is carried indirectly by the antioxidant capacity of other compounds to scavenge ROS and inhibit NO production, it is known that capsaicins and capsiates directly induce an anti-inflammatory effect through the activation of the transient receptor potential vanilloid 1 receptor (TRPV1) [24, 86]. TRPV1 is an ion channel receptor located in several glia that gives sensations of heat, but is also relevant in pain perception and induces this sensation, as well as inflammation [87]. By activating this receptor, capsaicins cause the well-known pungency sensation, as well as irritation in high enough concentrations. However, after activation by capsaicinoids, the excited neurons become resistant to further stimuli [87]. Hence, after a brief burning sensation, capsaicin may act as a local analgesic in neuropathic pain [84].

Research on capsaicinoids has showed that these molecules can also decrease inflammation of adipose tissue linked to obesity [73, 75]. It seems that capsaicin is able to inhibit or at least partially decrease the production of pro-inflammatory signals like interleukin 8 (IL-8), nuclear factor kappa-light-chain-enhancer of activated B cells (NF\kappaB) or active protein 1 (AP-1) [74, 80]. This anti-inflammatory effect has the potential to make inflamed tissue less prone to tumor development as well as reduce the infection caused inflammation [25].

Still, the pungency of capsaicins poses a setback for using them without potentially hazardous secondary effects. That is why capsinoids may be a good alternative to induce similar effects to capsaicin, even though they are not found in so many pepper species [41, 45].

Although the importance of researching capsaicinoids bioactivities in pepper is due to its unique presence in this genre, the also present phenolic compounds, vitamins and pigments also bear significant anti-inflammatory properties [25, 88, 89]. Indeed, the antioxidative properties of phenolic compounds could be a key factor in reducing age-related extended tissue inflammation [25, 58]. Nonetheless, further research assessing the suitability of using as anti-inflammatory drug these phytochemicals, and specifically capsaicin, is needed.

4. Extraction systems for bioactive compounds from *Capsicum* spp.

The bioactive compounds obtained from nature are secondary metabolites produced by the organisms. The concentrations of each type of compound are very variable, so in order to have enough quantities, the development of new and advanced technologies is needed.

Although there are several extraction methodologies, there is a demand for more appropriate and standardized extraction strategies. For choosing one method over another, it must be taken into account the quantitative and qualitative characteristics of the compound of interest. In addition, to improve the efficiency of a method it is important that the nature of the source, the different parameters of the method and the possible interaction are taken into account. Experiments carried out with peppers can be seen in **Table 6**.

4.1 Conventional extraction systems

4.1.1 Maceration

Maceration (MA) is a type of solid–liquid extraction, in which the solid would be peppers. This raw material is characterized by having several compounds soluble in the liquid phase. The type of molecules extracted will depend on two factors: the type of starting material and the type of solvent used.

In order to carry out this process, the sample is suspended in the solvent at the desired temperature during the chosen time while stirring. Once the process has been carried out, the sample is centrifuged and the supernatant, which is where the compounds of interest are found, is filtered. Therefore, the process of extraction consists of two stages. On the first (washing) there is a rapid transfer of the compounds from the surface of the solid to the solvent. On the second one (transfer), the matter passes from the interior to the exterior of the solid by diffusion

Compound	Species	Extraction		Medium	Yield	Ref.
		Type	Method			
Pigments	*Capsicum* spp.	Emergent	SFE	—	6%	[90]
	C. annuum	Green	MAE	Acetone-water	—	[91]
		Emergent	SFE	—	17.4%	[92]
				—	93%	[93]
				—	7.2%	[94]
Carotenoids	*C. annuum*	Green	EAE	Viscozyme L	87%	[95]
				Pectinase	80%	[95]
				Cellulase	Low	[95]
				Viscozyme L	78%	[96]
Phenolics	*C. annuum*	Emergent	SFE	—	84%	[93]
				—	100%	[94]
Capsaicinoids	*Capsicum* spp.	Green	MAE	Acetone (30%)	0.48 mg/g	[97]
	C. annuum	Green	MAE	None	230 ppm	[98]
			UAE	None	200 ppm	[97]
				n-hexane (100%)	—	[99]
				Methanol (100%)	100%	[100]
	C. annuum	Green	EAE	Viscozyme L	88.8%	[96]
				Viscozyme L	22%	[101]
				Celluclast	20%	[101]
		Emergent	SFE	—	8.6%	[92]
				—	—	[102]
				—	93%	[93]
				—	710 μg/g	[103]
	C. frutescens	Green	MAE	Acetone (pure)	5.3 mg/g	[104]
				Ethanol (99.5%)	—	[105]
			UAE	Ethanol (95%)	87.4%	[106]
				Methanol (100%)	100%	[100]
				Acetone (100%)	3.92 mg/g	[107]
		Emergent	SFE	—	710 μg/g	[103]
				—	5.2%	[108]
	C. chinense	Green	UAE	Acetone (100%)	0.31 mg/g	[109]
				Methanol (100%)	2.88 mg/g	[110]
		Emergent	SFE	—	0.5%	[111]
	C. baccatum	Green	UAE	Methanol (100%)	50%	[112]
Antioxidants	*C. baccatum*	Green	UAE	Methanol (100%)	27%	[112]
OLEORESIN	*C. baccatum*	Green	UAE	Methanol (100%)	26%	[112]
	C. annuum	Green	EAE	Viscozyme L	6%	[101]
		Emergent	SFE	—	8.2%	[102]
				—	7.4%	[94]
				—	—	[113]
	C. frutescens	Emergent	SFE	—	0.3%	[108]
Other phytochemicals	*Capsicum* spp.	Green	UAE	Methanol (80%)	—	[114]

Table 6.
Bioactive compounds obtained from Capsicum *spp. extracted with different extraction methods.*

being this step the limiting stage. This method is used, for example, for the isolation of phenolic compounds of vegetable origin for their subsequent use in the food, pharmaceutical and cosmetic industries [115].

Water can be used as an extraction solvent, but it comes with the drawback is that it principally extracts the hydrophilic compounds present in vegetable materials. Several studies showed that water is not a very suitable solvent because its high polarity does not allow the extraction of capsaicinoids that are non-polar. However, it has the advantages or being safe, cheap and the simplest form to obtain essential oils. This technique can also be done with other solvents, being the ratio of solvent another parameter to take into account. Polarity is the reason why the most common solvents used are methanol, ethanol, water or a mixture of them [3].

Another parameter to optimize will be temperature. This makes it possible to differentiate two types of MA: cold maceration and heat maceration. Maceration with or without stirring, mild heating or heating under reflux are also possible variations to the method.

It is a quite old simple method with the inconvenience of long extraction times and large amounts of sample and solvents. Additionally, high temperatures can destroy thermolabile compounds such as phenolic compounds. In a study in which the effect of MA and ultrasound was related, it was detected that they have similar extraction yield of oleoresin; nevertheless, MA needs longer extraction times with the consequent loss of quality due to their higher times of exposure to high temperatures. It also was observed that n-hexane was a better solvent than ethanol [116].

4.1.2 Heat assisted extraction

This procedure is also a type of solid-liquid extraction in which the extraction is usually carried out in thermostatic and sealed water baths so that the solvent does not evaporate. The determining parameters are extraction time, temperature and the proportion of solvent that is used as well as the solid-liquid ratio. Due to the application of high temperatures (as far as 100°C), it is not the best extraction process for this type of matrix because certain compounds present in peppers (vitamins, phenolic compounds, etc.) are degraded by heat.

4.1.3 Cold pressure extraction

Cold pressure extraction is one of the oldest techniques of extraction for obtaining oils. It consists of mechanical pressing with the absence of heating. By using this method, little to no heat is generated, however it gives low yields. This technique was applied to Capia pepper seed, resulting in a lower extraction yield compared to traditional Soxhlet extraction with hexane. Moreover, the final oil was unpopular among consumers [117].

4.1.4 Organic solvent extraction

Organic solvent extraction (OSE) allows the extraction of many compounds (oils, fats and proteins). Normally, after the extraction process another step is done. It consists of concentrating the extract by removing the solvent under atmospheric or reduced pressure. OSE is the most extensively used technique to obtain oleoresins from peppers. These oleoresins are. Usually used as color additives [118]. In this method, the most determining limitation will be the solvent properties (polarity). When choosing the solvent, it must also be taken into account if the legislation allows its use. In the EU is regulated in Commission Regulation N° 231/2012.

In order to improve these conventional extraction techniques, other methods have been developed (ultrasound assisted extraction, pressurized hot water extraction, negative pressure cavitation-assisted extraction and pulsed electric) to partner with them and solve some of the inopportuneness of traditional techniques. In this way the extraction efficiency is substantially improved.

4.2 Green extraction techniques

This group includes several techniques with the aim of not only preventing pollution but also reducing sample preparation costs by, for example, lowering solvents consumption.

4.2.1 Microwave-assisted extraction

Microwave-assisted extraction (MAE) is based on the heating of solvents that are in contact with solid samples with the use of microwaves. This allows the partition of compounds of interest from the sample into the solvent. Microwave energy accelerates a great variety of chemical reactions as well as the extraction of organic compounds from different matrices [98]. During the process, electromagnetic energy is transformed into calorific energy by two mechanisms: ionic conduction and dipole rotation [119]. It is a method that could be an alternative for avoiding thermal degradation and oxidation, with no influence on cell integrity and shape. However, due to thermal stress and localized high pressures, cell rupture is more rapid than with another techniques, which is an inconvenience, for example, in the extraction of volatile oils [91].

Some of the advantages of this technique are simplicity, effectiveness, low processing time, low solvent consumption and energy, no generation of secondary waste and can be used for larger volumes [120]. Another advantage of this method is that it produces a uniform heating, so extraction is simultaneous regardless of the area where the compound is [121]. By using this method, lipids, pigments, carbohydrates, vitamins and proteins can be extracted [122]. Moreover, it is characterized by having a superior extraction rate of volatile compounds in *Capsicum* spp.

The selectivity and efficiency of MAE largely depends on the dielectric constant of the extraction solvent mixture, which defines its chemical polarity and thus what compounds will be extracted [91]. It can be expected that by using water as a single solvent, the quantity of capsaicinoids extracted decrease due to its negligible polarity [109]. Therefore, the parameters to take into account in this type of extraction are time, power, temperature and type of solvent and its ratios.

Several experiments can be seen in a summarized form in **Table 7**, which shows the variation between the values of the parameters among different experiments with solvents of different polarity. It is worth highlighting the values obtained for capsaicinoids using ethanol [105] as solvents since it is the one with the highest yield.

4.2.2 Ultrasound assisted extraction

In every extraction process there is a critical step which is the degradation of cell walls and membranes. Ultrasound assisted extraction (UAE) makes this possible by applying pressure waves that are transmitted across a medium as compression and expansion (rarefaction) cycles, finding an area with maximum pressure in the compression phase and one with minimum pressure in the rarefaction phase. This pressure difference makes the cavitation phenomenon possible, which breaks the

Target	Species	Extracting conditions				Yield	Ref.
		Temp.	Time	Power	Solvent ratios		
		°C	min	W			
Volatiles	*Capsicum annuum*	55	10	250	n-hexane (pure)	42 compounds	[59]
Pigments	*Capsicum annuum*	60	2	50	Acetone-water (50%)	—	[91]
Capsaicinoids	*Capsicum* spp.	21	15	300	Acetone (30%)	0.48 mg/g	[98]
	Capsicum chinense	50	5	800	Methanol (100%)	0.32 mg/g	[109]
	Capsicum frutescens	—	20	320	Ethanol (99.5%)	5.3 mg/g	[105]
	Capsicum frutescens	120	15	150	Acetone (pure)	0.673%	[104]
	Capsicum annuum	76	1	500	Methanol (60%)	230 ppm	[97]

—, not described.

Table 7.
Different experimental conditions carried out in peppers by microwave-assisted extraction.

structure to make the compounds accessible and therefore available to extract them [123].

This technique has been applied to extract different bioactive compounds from herbs or algae due to its ease to disrupt their cell walls despite its resistance which differentiates this method from the previous ones, since strong disruption of the cell wall is achieved and as a result the extraction of intracellular materials is increased with the increment of energy input [106, 120, 124].

The ultrasonic technique, as well as the other green techniques, has been proven to have several advantages such as reduction of solvents consumption, temperature and time, very important parameters in the extraction of thermolabile and unstable compounds as it has no effect on the chemical structure and biological properties. Furthermore, UAE has low equipment investment and easy implementation, so it can be basically industrially employed in local companies [106]. Moreover, there are studies that showed that this method also achieves a greater supercritical extraction of pungent compounds from ginger owing to physical effects on the surface of particles [106].

There are several parameters to take into account to optimize the method which include ultrasound power intensities, frequency, wavelength and time. There are several studies about it (**Table 8**). One of them studied the influence of several of these parameters on the extraction of capsaicinoids, observing that their release was very fast in the first 5 min of the process and then decreased. Temperature influence was also observed since an increase of 15°C (up to 45°C) improved the extraction using 95% ethanol (v/v) as a solvent. Higher temperatures did not lead to significant improvements [106]. The effect of the solvent was also studied, being the most common solvents used for extracting capsaicinoids methanol, ethanol, acetone, acetonitrile and water. It was proved that the yield of extraction is worse with the addition of water. Methanol and ethanol have similar recoveries but ethanol was better than acetone, the best concentration of ethanol being 95% [100, 106].

Target	Species	Conditions					Yield	Ref.
		Freq.	Intensity	Time	Temp.	Solvent		
		kHz	W/cm^3	min	°C	(%)		
Capsaicinoids	*Capsicum chinense*	20	—	10	40	Acetone (100)	0.31 mg/g	[109]
	Capsicum chinense	40	—	10	24	Methanol (100)	2.88 mg/g	[110]
	Capsicum annuum	31.5	—	20	76	Methanol (60)	200 ppm	[97]
	Capsicum annuum	—	—	20	50	n-hexane (100)	—	[99]
	Capsicum annuum	20	360	10	50	Methanol (100)	100%	[100]
	Capsicum frutescens	35	600	180	45	Ethanol (95)	87.4%	[106]
	Capsicum frutescens	20	360	10	50	Methanol (100)	100%	[100]
	Capsicum frutescens	—	—	40	25	Acetone (100)	3.92 mg/g	[107]
	Capsicum baccatum	20	150	20	40	Methanol (100)	50%	[112]
Antioxidants	*Capsicum baccatum*	20	150	20	40	Methanol (100)	27%	[112]
Oleoresin	*Capsicum baccatum*	20	450	20	60	Methanol (100)	26%	[112]
Phytochemicals	*Capsicum*	35	—	20	50	Methanol (80)	—	[114]

—, not described.

Table 8.
Different experimental conditions carried out in peppers by ultrasound assisted extraction.

4.2.3 Enzyme-assisted extraction

Enzyme-assisted extraction (EAE) is an extraction system that allows the avoidance of processing conditions like temperature or drastic pH changes and so, maintain the quality and yield of multiple biomolecules [125].

This type of extraction supports isolation for recovering bioingredients from different plant materials. An enzymatic pre-treatment before applying traditional methods will help to isolate high yields of bioingredients due to enzyme assisted extraction facilities. This is due to the fact that degradation of cell walls and membranes is the critical step of extraction which [101]. Among its advantages over traditional methods are high selectivity, overall efficacy, eco-friendly procedures, low-energy consumption, minimal usage of harsh chemicals, maximum yield, low to no wasteful protection or deprotection steps, easy recovery, and process recyclability [126]. However, it also presents some drawbacks such as the cost of enzymes, requirement of holding tanks that may require long term incubation, lack of knowledge about optimal or compatible enzyme formulations for cell disruption and inability to completely hydrolyse the bonds in plant cell wall [127].

The factors to take into account in this technique are fundamentally temperature, pH, and type of enzyme. A range of enzymes (lipases, carbohydrase, celluloses, proteases, pectinases) have been widely used as specific catalysts. Each enzyme has different substrates. For example, cellulase and hemi-cellulase have their greater hydrolysing activity on the cellulose found in plant cell walls, hence their name. This enzymatic processing increases the permeability of the cell wall, resulting in a better recovery of some compounds like volatile oil and resin, which are prone to degradation when extracted with more disruptive methods [101].

Several studies (**Table 9**) show that viscozyme L is an enzyme with a superior recovery rate of bioactive compounds fractions like total carotenoid content, total phenolic content, total flavonoids and total antioxidant activity with high total suspended solids (TSS). It also has at this moment the better extract yield [95, 96]. Among its applications is the extraction of pigments or capsaicinoids. In almost all cases the pH of the medium is 4.5 as it is the pH of the sample and it is in the range of optimal activity of the different enzymes [95].

4.3 Emerging technologies for extraction

Due to the disadvantages of traditional techniques, there is an interest in the development of new extraction techniques. Some of the most sought-after features include shortened extraction time, automation or reduced organic solvent consumption.

4.3.1 Hydrostatic high-pressure extraction

Hydrostatic high pressure (HHP) is considered an emerging technology that have been applied in the preservation of food since the end of eighties [128]. The first food products treated with this method began commercialization in Japan in 1990 [129]. This method is established on the application of high pressures (100–900 MPa) to the product of interest.

Among the advantages of this type of extraction there is the possibility to conduct it at room temperature, meaning no thermal degradation and derived bioactivity loss from extracted components. It also does not modify chemical structures of the different compounds of interest's independent form of molecular weight [130]. Moreover, in comparison with conventional techniques it is faster,

Target	Enzyme	Conditions			Yield	Ref.
		Temp	pH	Time		
		°C		h	%	
Carotenoids	Viscozyme L	60	4.5	1	87	[95]
	Pectinase	60	4.5	1	80	[95]
	Cellulase	60	4.5	1	low	[95]
	Viscozyme L	50	4.5	5	78	[96]
Capsaicinoids	Viscozyme L	50	4.5	5	88.8	[96]
	Viscozyme L	45	4.5	1	22	[101]
	Celluclast	45	4.5	1	20	[101]
Oleoresin	Viscozyme L	45	4.5	1	6	[101]

Table 9.
Different experimental conditions carried out in C. annuum *by enzyme-assisted extraction.*

gives higher extraction yields and fewer impurities [131] giving, unlike other preservation technologies as thermal treatment, uniform and nearly instantaneous effects throughout the foodstuff and thus independent of foodstuff geometry and equipment size which makes an easy scale-up from laboratory findings to full-scale production possible [132].

This method can be applied to multiple matrices for the extraction of natural compounds [133] like fruit and vegetables for different targets as for example carotenoids [134], antioxidants [131] and pigments [135].

The main parameters to be taken into account are pressure, time, temperature and type and quantity of solvent, studying each parameter for each variety in particular.

HHP can change physiological and biochemical properties of pepper based on a study carried out with *C. annuum* [136]. In this research, different pressures (50, 100, 200 and 300 MPa) where applied for 5 min at 25°C observing that it produces remarkable changes in seedlings, but no or very little physiological and biochemical variations, creating an antioxidant system that is positive for the plant itself in the process. Therefore, HHP (500 MPa) can be used as a preservation treatment similar to pasteurization which was the ttraditional thermal treatment for sweet pepper preservation [137].

4.3.2 Supercritical-fluid extraction

In supercritical-fluid extraction (SFE), the fluid must reach temperature and pressure above the critical point, so as the fluid behaves like liquid and gas simultaneously which makes extraction easier. This technology has been used in a wide diversity of fields (food, pharmaceutical, chemical and fuel industries) due to its advantages, there is an absence of toxic residue in the final product which allows not only the extraction valuable active compounds (fatty acids, pigments, polyphenols and vitamins) free of solvents, but also to remove undesirable compounds (pollutants, toxins and pesticides) [138]. Additional advantages are great extraction selectivity, short processing times, requirement of minimal solvents, low degradability of the extracted product and the fact that the remaining biomass can be treated with other techniques in order to continue extraction.

The most important conditions are temperature, pressure and co-solvent. The selection of each parameter will depend on the specific compound searched [138]. The most used solvent is CO_2 due to its thermodynamics and heat transfer properties. Moreover, it has a low critical point (31°C, 73 bar). Furthermore, the polarity of CO_2 can be modified by the use of co-solvents such as ethanol, and in this way also extract polar components.

Studies demonstrate that the extracts obtained with this technique are better than natural spice for flavoring purposes as SFE could reduce aflatoxin in the final products [139]. In addition, numerous bibliographic references show that carotenoid extraction is better at higher pressures [113].

Furthermore, when analyzing the extraction of β-carotene and capsaicin at the same time, capsaicin shows lower solubility, yet the solubility of β-carotene did not change in the presence of capsaicin, a factor to bear in mind when designing the separation process of coloring and hot components from paprika [140]. Several pepper species have been used to obtain natural compounds of interest using supercritical fluid extraction including capsaicinoids, oleoresins, pigments, tocopherols and even aflatoxins (**Table 10**) observing large variations between the pressures applied in each of the methods referred to in the bibliography. The greatest amount of studies is related to the specie *C. annuum* since it is the best known, widespread and cultivated specie of the genus *Capsicum*.

Target	Species	Pressure	Temp.	Flow rate	Time	Pepper	Yield	Ref.
		MPa	°C	cm s^{-1}	min	g	%	
Capsaicinoids	*C. frutescens*	20.5	40	0.064	10	50	5.2	[108]
	C. frutescens	7.84 × 10^{-3}	55	1	—	70	7.1 × 10^{-4}	[103]
	C. annuum	55	40	—	—	—	8.6	[92]
	C. annuum	12	40	—	390	—	—	[102]
	C. annuum	40	55	0.9–1.2^a	—		93	[93]
	C. annuum	7.84 × 10^{-3}	55	1	—	250	7.1 × 10^{-4}	[103]
	C. chinense	15	60	2.15 × 10^{-3}	90	2.5	0.5	[111]
Oleoresins	*C. frutescens*	21.5	40	0.071	10	50	0.3	[108]
	C. annuum	30	40	—	360	—	8.2	[102]
	C. annuum	40	35	1–1.5	—		7.4	[94]
	C. annuum	43–54	40	1	—	25–30	—	[113]
Pigments	*Capsicum* spp.	47.5	80	sm	—	—	6	[90]
	C. annuum	36	45	—	—	—	17.4	[92]
	C. annuum	20	35	0.9–1.2	—	—	93	[93]
	C. annuum	40	35	1–1.5	—	—	7.2	[94]
Aflatoxins	*C. annuum*	30	50	—	2	7.8	16.2	[139]
Tocopherols	*C. annuum*	5	25	0.9–1.2^a	—	—	84	[93]
	C. annuum	40	35	1–1.5	—	—	100	[94]

sm, semi-continuous; —, not described.
[a]Propane.

Table 10.
Different experimental conditions carried out in peppers by supercritical-fluid extraction.

4.3.3 Pulse electric field extraction

Pulsed electric field (PEF) consists of a non-thermal method that is extensively used in food processing applications due to, among other things, its ability to kill microorganisms in liquid foods. This method involves the application of short duration electric pulses, making several pores on the cell membrane in what is called electropermeabilization or electroporation. This makes it possible for the selective recovery of intracellular components with low energy consumption as the dielectric breakdown theory explains. According to this theory, the membrane of the cell has a low-dielectric constant that when exposed to a strong electric field provokes ion migration, forming free charges of the opposite sign which accumulate at both membrane sides, generating a potential difference across the membrane which depends on the size and shape of the cells and the concentration of cells in suspension. This difference in charges makes cell walls undergo a compression, reducing the membrane thickness that results in the formation of micropores and increasing permeability (electroporation) [119, 141]. It can be used directly or as a pre-treatment prior to solvent extraction [142]. The parameters to take into account with this method are number of pulses, length of the pulses, energy input and biomass concentration.

Among the advantages of PEF is the short extraction time (usually under 1 s), efficiency at low temperatures, decreased energy losses and a successful cell wall breakdown [120]. The disadvantages are that membrane changes can be reversible, air bubbles make the process less effective and the efficiency of the method depends on electric field strength and electrode gap [142].

Further research is necessary since there is no data on the extraction of compounds with this technique. However, there are studies of the application of PEF for the production of dehydrated products or juices. As for drying, a study [143] done with red bell pepper at 320 J/kg, 2.0 kV/cm, 1 Hz and 30 pulses of 400 μs reported good results. Another study [144] with the same aim used different extraction conditions, which were 2.5 kV/cm, 100 Hz and 30 μs as the pulse time. As for juice production, *C. annuum* was studied with extracting conditions of 500 J/kg, 1.7 kV/cm, 1 Hz and 30 pulses of 300 μs demonstrating that juice from PEF treated paprika compared well in quality with enzymatic treatment [145].

5. Seed valorisation and industrial application

As it is being already done by some factories and companies in pepper growing countries, seeds, stalks or peels are extracted through whole fruit processing with or without semi-automated equipment. This is mainly done by the food industry in order to take away the non-edible parts of the vegetable to elaborate commodities as pepper powders (e.g., paprika), sauces, jams or pickled peppers. This way, stalks, seeds and placenta are usually removed and discarded.

But in many cases, placenta and seeds are subjected to extraction techniques in order to obtain essential oils to treat chronic ailments via topical administration in traditional remedies [146]. Even so, it is clear that pepper seeds are a multivalent source of many important compounds that could be used in different applications.

The elevated phenolic concentrations found in seeds make them an excellent target for extraction and use of these phenolic compounds, whether it is as antioxidant supplements, as food coating in order to increase shelf-life or as a potential source of functional animal feed. Moreover, the implementation of an extraction process in the food industry is dependent on the type of compounds of interest, as the yields of phytochemicals obtained may differ greatly. It should be noted that highly efficient and sustainable extraction methods continue to be a focus of research both in industry and academia [65].

Also, antimicrobial potential of phytochemicals present in pepper should be considered. Even though this would involve highly specialized equipment and maintain strict aseptic conditions, there is promise in their use as food preservers from common food-borne pathogenic microorganisms besides their antioxidative capability. Furthermore, their use as alternative antibiotics has hardly been researched and the safety of their administration or clinical use has not yet been assessed.

Regarding their nutrition attributes, the high linoleic acid and palmitic acid present in the seeds could lead to use them as a primary source of cooking oil, or as a main component in foods such as margarine, or preserver in certain pickled products [48]. Also, as they are a rich natural source of vitamins E and C, extracts of these vitamins could also be incorporated as constituents of many juices and other foods. Pepper-processing industry could find very rewarding to investigate on the yet unknown beneficial properties of its by-products and obtain additional profit by integrating existing extraction and purification technologies into their process chain.

6. Conclusions

It is clear that the wide spectrum of historical applications of peppers is supported by the latest findings on the properties of peppers. The diversity among

Capsicum species properties have been studied over the years, although many of the possible bioactivities have only been characterized in the whole fruit and there are few *in vivo* experiments reported testing phytochemicals of this genus or assessing its safety. Moreover, research on produce by-products has acquired weight within the scientific community and industry as of lately. The exploitation of food processing by-products means not only an affordable raw material for the extraction of valuable phytochemicals, but also a reduction of ecological and economic impact of agricultural and industrial activities. Further studies are necessary to explore in depth more efficient bio-active compounds from the underutilized and non-edible biomass of *Capsicum* spp. Furthermore, the great genotypic diversity of the many pepper varieties grown in different regions, have a high variety of bioactive compounds, some still to be discovered. Currently available knowledge about the medicinal phytochemicals found in the genus *Capsicum* should be considered to develop novel drugs to treat many diseases and ailments.

Acknowledgements

Many thanks to MICINN for the financial support for the Ramón & Cajal researcher of M.A. Prieto, to the Regional Government of Galicia to the Regional Government of Galicia for "Programa de axudas á etapa predoutoral da Xunta de Galicia" for the pre-doctoral researchers of A.G. Pereira and the authors are also grateful to Interreg España-Portugal for financial support through the 0377_Iberphenol_ project.

Author details

Javier Echave[1], Antía G. Pereira[1,2], María Carpena[1], Miguel Ángel Prieto[1*] and Jesús Simal-Gandara[1*]

1 Nutrition and Bromatology Group, Analytical and Food Chemistry Department, Faculty of Food Science and Technology, University of Vigo, Ourense, Spain

2 Centro de Investigação de Montanha (CIMO), Instituto Politécnico de Bragança, Bragança, Portugal

*Address all correspondence to: mprieto@uvigo.es and jsimal@uvigo.es

References

[1] Pérez-castañeda M, Castañón-nájera G, Ramírez-meraz M. Avances y perspectivas sobre el estudio del origen y la diversidad genética de *Capsicum* spp. Ecosistemas y Recursos Agropecuarios. 2015;**2**:117-128

[2] Simon JE, Chadwick AF, Craker LE. Herbs: An Indexed Bibliography 1971–1980: The Scientific Literature on Selected Herbs, and Aromatic and Medicinal Plants of the Temperate Zone. Hamden, CT: Archon Books; 1984

[3] Melgar-Lalanne G, Hernández-Álvarez AJ, Jiménez-Fernández M, Azuara E. Oleoresins from *Capsicum* spp.: Extraction methods and bioactivity. Food and Bioprocess Technology. 2017;**10**:51-76

[4] Pickersgill B. Migrations of chili peppers, *Capsicum* spp., in the Americas. In: Pre-Columbian Plant Migration. Papers of the Peabody Museum of Archaeology and Ethnology. 1984. pp. 105-123

[5] Kraft KH, Brown CH, Nabhan GP, Luedeling E, De Jesús Luna Ruiz J, D'Eeckenbrugge GC, et al. Multiple lines of evidence for the origin of domesticated chili pepper, *Capsicum annuum*, in Mexico. Proceedings of the National Academy of Sciences of the United States of America. 2014;**111**:6165-6170

[6] Castañon-Najera G, Mayek-Pérez N, Ruiz-Salazar R, Garcia AC. Molecular study of a diallel chilli with amplified fragment length polymorphism (AFLP) markers. African Journal of Agricultural Research. 2011;**6**:6126-6131

[7] Hui YH, Chen F, Nollet LML, Guiné RPF, Le Quéré JL, Martín-Belloso O, et al. Handbook of Fruit and Vegetable Flavors. Hoboken, New Jersey: John Wiley and Sons Inc.; 2010

[8] Basu SK, De AK. Capsicum: Historial and Botanical Perspectives in *Capsicum*, The genus *Capsicum*. London, UK: CRC Press; 2003. pp. 1-15

[9] Martín NC, González WG. Caracterización de accesiones de chile (*Capsicum* spp.). Agronomía Mesoamericana. 2016;**2**:31

[10] Barrios O, Fuentes V, Abreu S. Especies cultivadas de ajies y pimientos (*Capsicum* spp. div.) en Cuba. In: Proceedings of the Convención Trópico 2004. II Congreso de Agricultura Tropical. 2004. pp. 1-15

[11] Castañón-Nájera G, Latournerie-Moreno L, Lesher-Gordillo JM, De La Cruz-Lázaro E, Mendoza-Elos M. Identification of variables for the morphological characterisation of hot peppers (*Capsicum* spp) in Tabasco, Mexico. Universidad y Ciencia. 2010;**26** (3):225-234

[12] Katz E. Chili Pepper, from Mexico to Europe: Food, imaginary and cultural identity. In: Estudios del Hombre, Serie Antropología de la Alimentación. Food, Imaginaries and Cultural Frontiers. Essays in Honour of Helen Macbeth. Guadalajara, Mexico: Universidad de Guadalajara; 2009. pp. 213-232

[13] Rajput J, Parulekar Y. El pimiento. In: Tratado de Ciencia y Tecnología de las hortalizas. Zaragoza: Acriba; 2004. pp. 203-225

[14] Martínez S, Curros A, Bermúdez J, Carballo J, Franco I. The composition of Arnoia peppers (*Capsicum annuum* L.) at different stages of maturity. International Journal of Food Sciences and Nutrition. 2007;**58**:150-161

[15] Orellana-Escobedo L, Garcia-Amezquita LE, Olivas GI, Ornelas-Paz JJ, Sepulveda DR. Capsaicinoids content and proximate composition of Mexican chili peppers (*Capsicum* spp.) cultivated in the State of Chihuahua. CyTA—Journal of Food. 2013;**11**:179-184

[16] Zou Y, Ma K, Tian M. Chemical composition and nutritive value of hot pepper seed (*Capsicum annuum*) grown in northeast region of China. Journal of Food Science and Technology. 2015; **35**:659-663

[17] Esayas K, Shimelis A, Ashebir F, Negussie R, Tilahun B, Gulelat D. Proximate composition, mineral content and antinutritional factors of some capsicum (*Capsicum annum*) varieties grown in Ethiopia. Bulletin of the Chemical Society of Ethiopia. 2011;**25**: 451-454

[18] Ku K-H, Choi E-J, Park JB. Chemical component analysis of red pepper (*Capsicum annuum* L.) seeds with various cultivars. Journal of the Korean Society of Food Science and Nutrition. 2008;**37**:1084-1089

[19] Liu S, Li W, Wu Y, Chen C, De Lei J. Novo transcriptome assembly in chili pepper (*Capsicum frutescens*) to identify genes involved in the biosynthesis of Capsaicinoids. PLoS One. 2013;**8**:1-8

[20] Arimboor R, Natarajan RB, Menon KR, Chandrasekhar LP, Moorkoth V. Red pepper (*Capsicum annuum*) carotenoids as a source of natural food colors: Analysis and stability—A review. Journal of Food Science and Technology. 2015;**52**: 1258-1271

[21] Ciulu-Costinescu F, Chifiriuc MC, Popa M, Bleotu C, Neamtu J, Marina L, et al. Screening of polyphenol content and in vitro studies of antioxidant, antibacterial and cytotoxic activities of *Capsicum annuum* extracts. Revista de Chimie—Bucharest. 2015;**66**:1261-1266

[22] Hernández-Ortega M, Ortiz-Moreno A, Hernández-Navarro MD, Chamorro-Cevallos G, Dorantes-Alvarez L, Necoechea-Mondragón H. Antioxidant, antinociceptive, and anti-inflammatory effects of carotenoids extracted from dried pepper (*Capsicum annuum* L.). Journal of Biomedicine and Biotechnology. 2012;**2012**:10. Article ID 524019

[23] Srinivasan K. Biological activities of red pepper (*Capsicum annuum*) and its pungent principle capsaicin: A review. Critical Reviews in Food Science and Nutrition. 2016;**56**:1488-1500

[24] Adaszek Ł, Gadomska D, Mazurek Ł, Łyp P, Madany J, Winiarczyk S. Properties of capsaicin and its utility in veterinary and human medicine. Research in Veterinary Science. 2019; **123**:14-19

[25] Pandey MK, Gupta SC, Nabavizadeh A, Aggarwal BB. Regulation of cell signaling pathways by dietary agents for cancer prevention and treatment. Seminars in Cancer Biology. 2017;**46**:158-181

[26] Rodrigues CA, Nicácio AE, Jardim ICSF, Visentainer JV, Maldaner L. Determination of phenolic compounds in red sweet pepper (*Capsicum annuum* L.) using using a modified QuEChERS method and UHPLC-MS/MS Analysis and its relation to antioxidant activity. Journal of the Brazilian Chemical Society. 2019;**30**: 1229-1240

[27] Iorizzi M, Lanzotti V, Giancarlo R, De Marino S, Zollo F. Antimicrobial Furostanol Saponins from the seeds of *Capsicum annuum* L. Var. *acuminatum*. Journal of Agricultural and Food Chemistry. 2002;**50**:4310-4316

[28] Kuku A, Odekanyin O, Adeniran K, Adewusi M, Olonade T, et al. Purification of a mannose/glucose-specific lectin with antifungal activity from pepper seeds (*Capsicum annum*). African Journal of Biochemistry Research. 2009;**3**:272-278

[29] Silva LR, Azevedo J, Pereira MJ, Valentão P, Andrade PB. Chemical assessment and antioxidant capacity of

pepper (*Capsicum annuum* L.) seeds. Food and Chemical Toxicology. 2013;**53**: 240-248

[30] Sandoval-Castro CJ, Valdez-Morales M, Oomah BD, Gutiérrez-Dorado R, Medina-Godoy S, Espinosa-Alonso LG. Bioactive compounds and antioxidant activity in scalded Jalapeño pepper industrial byproduct (*Capsicum annuum*). Journal of Food Science and Technology. 2017;**54**:1999-2010

[31] Nascimento PLA, Nascimento TCES, Ramos NSM, Silva GR, Gomes JEG, Falcão REA, et al. Quantification, antioxidant and antimicrobial activity of phenolics isolated from different extracts of *Capsicum frutescens* (Pimenta Malagueta). Molecules. 2014;**19**: 5434-5447

[32] Diz MSS, Carvalho AO, Rodrigues R, Neves-Ferreira AGC, da Cunha M, Alves EW, et al. Antimicrobial peptides from chilli pepper seeds causes yeast plasma membrane permeabilization and inhibits the acidification of the medium by yeast cells. Biochimica et Biophysica Acta (BBA)—General Subjects. 1760; **2006**:1323-1332

[33] Aguirre J, Ríos-Momberg M, Hewitt D, Hansberg W. Reactive oxygen species and development in microbial eukaryotes. Trends in Microbiology. 2005;**13**:111-118

[34] Chamikara MDM, Dissanayake DRRP, Ishan M, Sooriyapathirana SDSS. Dietary, anticancer and medicinal properties of the phytochemicals in chili pepper (*Capsicum* spp.). Ceylon Journal of Science. 2016;**3**:5-20

[35] Mokhtar M, Ginestra G, Youcefi F, Filocamo A, Bisignano C, Riazi A. Antimicrobial activity of selected polyphenols and capsaicinoids identified in pepper (*Capsicum annuum* L.) and their possible mode of interaction. Current Microbiology. 2017;**74**: 1253-1260

[36] Badhani B, Sharma N, Kakkar R. Gallic acid: A versatile antioxidant with promising therapeutic and industrial applications. RSC Advances. 2015;**5**: 27540-27557

[37] Kevers C, Falkowski M, Tabart J, Defraigne JO, Dommes J, Pincemail J. Evolution of antioxidant capacity during storage of selected fruits and vegetables. Journal of Agricultural and Food Chemistry. 2007;**55**:8596-8603

[38] Bertão MR, Moraes MC, Palmieri DA, Silva LP, da Silva RMG. Cytotoxicity, genotoxicity and antioxidant activity of extracts from *Capsicum* spp. Research Journal of Medicinal Plant. 2016;**10**: 265-275

[39] Sim KH, Sil HY. Antioxidant activities of red pepper (*Capsicum annuum*) pericarp and seed extracts. International Journal of Food Science and Technology. 2008;**43**:1813-1823

[40] Yahfoufi N, Alsadi N, Jambi M, Matar C. The immunomodulatory and anti-inflammatory role of polyphenols. Nutrients. 2018;**10**:1-23

[41] Aza-González C, Núñez-Palenius HG, Ochoa-Alejo N. Molecular biology of capsaicinoid biosynthesis in chili pepper (*Capsicum* spp.). Plant Cell Reports. 2011;**30**:695-706

[42] Chouaibi M, Rezig L, Hamdi S, Ferrari G. Chemical characteristics and compositions of red pepper seed oils extracted by different methods. Industrial Crops and Products. 2019;**128**: 363-370

[43] Howard LR, Talcott ST, Brenes CH, Villalon B. Changes in phytochemical and antioxidant activity of selected pepper cultivars (*Capsicum* species) as

influenced by maturity. Journal of Agricultural and Food Chemistry. 2000; **48**:1713-1720

[44] Deepa N, Kaur C, George B, Singh B, Kapoor HC. Antioxidant constituents in some sweet pepper (*Capsicum annuum* L.) genotypes during maturity. LWT—Food Science and Technology. 2007;**40**: 121-129

[45] Kobata K, Todo T, Yazawa S, Iwai K, Watanabe T. Novel capsaicinoid-like substances, capsiate and dihydrocapsiate, from the fruits of a nonpungent cultivar, CH-19 sweet, of pepper (*Capsicum annuum* L.). Journal of Agricultural and Food Chemistry. 1998;**46**:1695-1697

[46] Luo XJ, Peng J, Li YJ. Recent advances in the study on capsaicinoids and capsinoids. European Journal of Pharmacology. 2011;**650**:1-7

[47] Ludy MJ, Moore GE, Mattes RD. The effects of capsaicin and capsiate on energy balance: Critical review and meta-analyses of studies in humans. Chemical Senses. 2012;**37**:103-121

[48] Jarret RL, Levy IJ, Potter TL, Cermak SC. Seed oil and fatty acid composition in *Capsicum* spp. Journal of Food Composition and Analysis. 2013; **30**:102-108

[49] Matthäus B, Özcan MM. Chemical evaluation of some paprika (*Capsicum annuum* L.) seed oils. European Journal of Lipid Science and Technology. 2009; **111**:1249-1254

[50] Ristić-Medić D, Vučić V, Takić M, Karadžić I, Glibetić M. Polyunsaturated fatty acids in health and disease. Journal of the Serbian Chemical Society. 2013; **78**:1269-1289

[51] Materska M, Perucka I. Antioxidant activity of the main phenolic compounds isolated from hot pepper fruit (*Capsicum annuum* L.). Journal of Agricultural and Food Chemistry. 2005;**53**:1750-1756

[52] Chu YF, Sun J, Wu X, Liu RH. Antioxidant and antiproliferative activities of common vegetables. Journal of Agricultural and Food Chemistry. 2002;**50**:6910-6916

[53] Hernández-Ortega M, Ortiz-Moreno A, Hernández-Navarro MD, Chamorro-Cevallos G, Dorantes-Alvarez L, Necoechea-Mondragón H, et al. Antioxidant, antinociceptive, and anti-inflammatory effects of carotenoids extracted from dried pepper (*Capsicum annuum* L.). Journal of Biomedicine & Biotechnology. 2012;**2012**:1-10

[54] Rao AV, Rao LG. Carotenoids and human health. Pharmacological Research. 2007;**55**:207-216

[55] Mocchegiani E, Costarelli L, Giacconi R, Malavolta M, Basso A, Piacenza F, et al. Vitamin E-gene interactions in aging and inflammatory age-related diseases: Implications for treatment. A systematic review. Ageing Research Reviews. 2014;**14**:81-101

[56] Saini RK, Keum YS. GC-MS and HPLC-DAD analysis of fatty acids and tocopherols in sweet peppers (*Capsicum annuum* L.). Journal of Food Measurement and Characterization. 2016;**10**:685-689

[57] Asnin L, Park SW. Isolation and analysis of bioactive compounds in capsicum peppers. Critical Reviews in Food Science and Nutrition. 2015;**55**: 254-289

[58] Shahidi F, de Camargo AC. Tocopherols and tocotrienols in common and emerging dietary sources: Occurrence, applications, and health benefits. International Journal of Molecular Sciences. 2016;**17**:1745

[59] Gogus F, Ozel MZ, Keskin H, Yanik DK, Lewis AC. Volatiles of fresh and commercial sweet red pepper

pastes: Processing methods and microwave assisted extraction. International Journal of Food Properties. 2015;**18**:1625-1634

[60] De Lucca AJ, Bland JM, Vigo CB, Cushion M, Selitrennikoff CP, Peter J, et al. CAY-1, a fungicidal saponin from *Capsicum* sp. fruit. Medical Mycology. 2002;**40**:131-137

[61] Simonovska J, Škerget M, Knez Ž, Srbinoska M, Kavrakovski Z, Grozdanov A, et al. Physicochemical characterization and bioactive compounds of stalk from hot fruits of *Capsicum annuum* L. Macedonian Journal of Chemistry and Chemical Engineering. 2016;**35**:199-208

[62] Jang H-W, Ka M-H, Lee K-G. Antioxidant activity and characterization of volatile extracts of *Capsicum annuum* L. and *Allium* spp. Flavour and Fragrance Journal. 2008;**23**: 178-184

[63] Cichewicz RH, Thorpe PA. The antimicrobial properties of Chile peppers (*Capsicum* species) and their uses in Mayan medicine. Journal of Ethnopharmacology. 1996;**52**:61-70

[64] Von Borowski RG, Barros MP, da Silva DB, Lopes NP, Zimmer KR, Staats CC, et al. Red pepper peptide coatings control *Staphylococcus epidermidis* adhesion and biofilm formation. International Journal of Pharmaceutics. 2020;**574**:118872

[65] de Camargo AC, Schwember AR, Parada R, Garcia S, Maróstica Júnior MR, Franchin M, et al. Opinion on the hurdles and potential health benefits in value-added use of plant food processing by-products as sources of phenolic compounds. International Journal of Molecular Sciences. 2018:19

[66] Shahidi F, Ambigaipalan P. Phenolics and polyphenolics in foods, beverages and spices: Antioxidant activity and health effects—A review. Journal of Functional Foods. 2015;**18**: 820-897

[67] Azabou S, Ben Taheur F, Jridi M, Bouaziz M, Nasri M. Discarded seeds from red pepper (*Capsicum annum*) processing industry as a sustainable source of high added-value compounds and edible oil. Environmental Science and Pollution Research. 2017;**24**: 22196-22203

[68] Sora GTS, Haminiuk CWI, da Silva MV, Zielinski AAF, Gonçalves GA, Bracht A, et al. A comparative study of the capsaicinoid and phenolic contents and in vitro antioxidant activities of the peppers of the genus *Capsicum*: An application of chemometrics. Journal of Food Science and Technology. 2015;**52**: 8086-8094

[69] Hallmann E, Rembial Kowska E. Characterisation of antioxidant compounds in sweet bell pepper (*Capsicum annuum* L.) under organic and conventional growing systems. Journal of the Science of Food and Agriculture. 2012;**92**:2409-2415

[70] Dorantes L, Colmenero R, Hernandez H, Mota L, Jaramillo ME, Fernandez E, et al. Inhibition of growth of some foodborne pathogenic bacteria by *Capsicum annum* extracts. International Journal of Food Microbiology. 2000;**57**:125-128

[71] Bacon K, Boyer R, Denbow C, O'Keefe S, Neilson A, Williams R. Antibacterial activity of jalapeño pepper (*Capsicum annuum* var. *annuum*) extract fractions against select foodborne pathogens. Food Science & Nutrition. 2017;**5**:730-738

[72] Von Borowski RG, Zimmer KR, Leonardi BF, Trentin DS, Silva RC, de Barros MP, et al. Red pepper *Capsicum baccatum*: Source of antiadhesive and antibiofilm compounds against nosocomial bacteria. Industrial Crops and Products. 2019;**127**:148-157

[73] Kang JH, Kim CS, Han IS, Kawada T, Yu R. Capsaicin, a spicy component of hot peppers, modulates adipokine gene expression and protein release from obese-mouse adipose tissues and isolated adipocytes, and suppresses the inflammatory responses of adipose tissue macrophages. FEBS Letters. 2007;**581**:4389-4396

[74] Sancho R, Lucena C, Macho A, Calzado MA, Blanco-Molina M, Minassi A, et al. Immunosuppressive activity of capsaicinoids: Capsiate derived from sweet peppers inhibits NF-κB activation and is a potent antiinflammatory compound in vivo. European Journal of Immunology. 2002; **32**:1753-1763

[75] Choi S-E, Kim TH, Yi S-A, Hwang YC, Hwang WS, Choe SJ, et al. Capsaicin attenuates palmitate-induced expression of macrophage inflammatory protein 1 and interleukin 8 by increasing palmitate oxidation and reducing c-Jun activation in THP-1 (human acute monocytic leukemia cell) cells. Nutrition Research. 2011;**31**:468-478

[76] Metcalfe NB, Alonso-Alvarez C. Oxidative stress as a life-history constraint: The role of reactive oxygen species in shaping phenotypes from conception to death. Functional Ecology. 2010;**24**:984-996

[77] Rao V, Al-Weshahy A. Plant-based diets and control of lipids and coronary heart disease risk. Current Atherosclerosis Reports. 2008;**10**: 478-485

[78] Tundis R, Loizzo MR, Menichini F, Bonesi M, Conforti F, Statti G, et al. Comparative study on the chemical composition, antioxidant properties and hypoglycaemic activities of two *Capsicum annuum* L. cultivars (Acuminatum small and Cerasiferum). Plant Foods for Human Nutrition. 2011; **66**:261-269

[79] Marini E, Magi G, Mingoia M, Pugnaloni A, Facinelli B. Antimicrobial and anti-virulence activity of capsaicin against erythromycin-resistant, cell-invasive group A streptococci. Frontiers in Microbiology. 2015;**6**:1-7

[80] Lee IO, Lee KH, Pyo JH, Kim JH, Choi YJ, Lee YC. Anti-inflammatory effect of capsaicin in *Helicobacter pylori*-infected gastric epithelial cells. Helicobacter. 2007;**12**:510-517

[81] Yamasaki S, Asakura M, Neogi SB, Hinenoya A, Iwaoka E, Aoki S. Inhibition of virulence potential of *Vibrio cholerae* by natural compounds. The Indian Journal of Medical Research. 2011;**133**:232

[82] De Lucca AJ, Boue S, Sien T, Cleveland TE, Walsh TJ. Silver enhances the in vitro antifungal activity of the saponin, CAY-1. Mycoses. 2011;**54**:e1-e9

[83] Bashyal P, Parajuli P, Pandey RP, Sohng JK. Microbial biosynthesis of antibacterial chrysoeriol in recombinant *Escherichia coli* and bioactivity assessment. Catalysts. 2019;**9**(2):1-15

[84] Salehi B, Hernández-Álvarez AJ, Contreras MDM, Martorell M, Ramírez-Alarcón K, Melgar-Lalanne G, et al. Potential phytopharmacy and food applications of *Capsicum* spp.: A comprehensive review. Natural Product Communications. 2018;**13**: 1543-1556

[85] Ng TB, Ngai PHK, Cheung RCF, Wong JH, Lam SK, Wang HX, et al. Antifungal and mitogenic activities of cluster pepper (*Capsicum frutescens*) seeds. Nuts and Seeds in Health and Disease Prevention. 2011:345-349

[86] Malek N, Mrugala M, Makuch W, Kolosowska N, Przewlocka B, Binkowski M, et al. A multi-target approach for pain treatment: Dual inhibition of fatty acid amide hydrolase

and TRPV1 in a rat model of osteoarthritis. Pain. 2015;**156**:890-903

[87] Du Q, Liao Q, Chen C, Yang X, Xie R, Xu J. The role of transient receptor potential Vanilloid 1 in common diseases of the digestive tract and the cardiovascular and respiratory system. Frontiers in Physiology. 2019;**10** (1064):1-17

[88] Deng Y, Huang X, Wu H, Zhao M, Lu Q, Israeli E, et al. Some like it hot: The emerging role of spicy food (capsaicin) in autoimmune diseases. Autoimmunity Reviews. 2016;**15**: 451-456

[89] Sharma SK, Vij AS, Sharma M. Mechanisms and clinical uses of capsaicin. European Journal of Pharmacology. 2013;**720**:55-62

[90] Skerget M, Knez Z. Modeling high pressure extraction process. Computers & Chemical Engineering. 2001;**25**: 879-886

[91] Csiktusnádi Kiss GA, Forgács E, Cserháti T, Mota T, Morais H, Ramos A. Optimisation of the microwave-assisted extraction of pigments from paprika (*Capsicum annuum* L.) powders. Journal of Chromatography A. 2000;**889**:41-49

[92] Fernández-Trujillo JP. Supercritical CO_2 extraction of sweet and hot paprika oleoresin and other fractions. Grasas y Aceites. 2008;**59**(1):7-15

[93] Gnayfeed MH, Daood HG, Illés V, Biacs PA. Supercritical CO_2 and subcritical propane extraction of pungent paprika and quantification of carotenoids, tocopherols, and capsaicinoids. Journal of Agricultural and Food Chemistry. 2001;**49**:2761-2766

[94] Illés V, Daood HG, Biacs PA, Gnayfeed MH, Mészáros B. Supercritical CO_2 and subcritical propane extraction of spice red pepper oil with special regard to carotenoid and tocopherol content. Journal of Chromatographic Science. 1999;**37**:345-352

[95] Nath P, Kaur C, Rudra SG, Varghese E. Enzyme-assisted extraction of carotenoid-rich extract from red capsicum (*Capsicum annuum*). Agricultural Research. 2016;**5**:193-204

[96] Santamaría RI, Reyes-Duarte MD, Bárzana E, Fernando D, Gama FM, Mota M, et al. Selective enzyme-mediated extraction of capsaicinoids and carotenoids from chili guajillo puya (*Capsicum annuum* L.) using ethanol as solvent. Journal of Agricultural and Food Chemistry. 2000;**48**:3063-3067

[97] Paduano A, Caporaso N, Santini A, Sacchi R. Microwave and ultrasound-assisted extraction of capsaicinoids from chili peppers (*Capsicum annuum* L.) in flavored olive oil. Journal of Food Research. 2014;**3**:51

[98] Williams OJ, Raghavan GSV, Orsat V, Dai J. Microwave-assisted extraction of capsaicinoids from capsicum fruit. Journal of Food Biochemistry. 2004;**28**:113-122

[99] Rafajlovska V, Klopcevska J, Klopcevska J, Srbinoska M, Srbinoska M, Raicki SR. Ultrasound-assisted extraction of capsaicinoids and carotenoids from hot red pepper. In: Proceedings of the 5th Black Sea Basin Conference of Analytical Chemistry. 2009

[100] Barbero GF, Liazid A, Palma M, Barroso CG. Ultrasound-assisted extraction of capsaicinoids from peppers. Talanta. 2008;**75**:1332-1337

[101] Baby KC, Ranganathan TV. Effect of enzyme pretreatment on yield and quality of fresh green chilli (*Capsicum annuum* L) oleoresin and its major capsaicinoids. Biocatalysis and Agricultural Biotechnology. 2016;**7**: 95-101

[102] Govindarajan VS. Capsicum—Production, technology, chemistry, and quality—Part II. Processed products, standards, world production and trade. Critical Reviews in Food Science and Nutrition. 1986;**23**: 207-288

[103] Peusch M, Müller-Seitz E, Petz M, Müller A, Anklam E. Extraction of capsaicinoids from chillies (*Capsicum frutescens* L.) and paprika (*Capsicum annuum* L.) using supercritical fluids and organic solvents. European Food Research and Technology. 1997;**204**:351-355

[104] Nazari F, Ebrahimi SN, Talebi M, Rassouli A, Bijanzadeh HR. Multivariate optimisation of microwave-assisted extraction of capsaicin from *Capsicum frutescens* L. and quantitative analysis by 1H-NMR. Phytochemical analysis. 2007; **18**:333-340

[105] Chuichulcherm S, Prommakort S, Srinophakun P, Thanapimmetha A. Optimization of capsaicin purification from *Capsicum frutescens* Linn. with column chromatography using Taguchi design. Industrial Crops and Products. 2013;**44**:473-479

[106] Boonkird S, Phisalaphong C, Phisalaphong M. Ultrasound-assisted extraction of capsaicinoids from *Capsicum frutescens* on a lab- and pilot-plant scale. Ultrasonics Sonochemistry. 2008;**15**:1075-1079

[107] Deng X-Y. Optimization of ultrasonic-assisted extraction procedure of capsaicinoids from chili peppers using orthogonal array experimental design. African Journal of Biotechnology. 2012;**11**:13153-13161

[108] Duarte C, Moldão-Martins M, Gouveia AF, da Costa SB, Leitão AE, Bernardo-Gil MG. Supercritical fluid extraction of red pepper (*Capsicum frutescens* L.). Journal of Supercritical Fluids. 2004;**30**:155-161

[109] Vázquez-Espinosa M, De Peredo AVG, Ferreiro-González M, Barroso CG, Palma M, Barbero GF, et al. Optimizing and comparing ultrasound- and microwave-assisted extraction methods applied to the extraction of antioxidant capsinoids in peppers. Agronomy. 2019;**9**(633):1-18

[110] Sganzerla M, Coutinho JP, de Melo AMT, Godoy HT. Fast method for capsaicinoids analysis from *Capsicum chinense* fruits. Food Research International. 2014;**64**:718-725

[111] de Aguiar AC, dos Santos P, Coutinho JP, Barbero GF, Godoy HT, Martínez J. Supercritical fluid extraction and low pressure extraction of Biquinho pepper (*Capsicum chinense*). LWT—Food Science and Technology. 2014;**59**: 1239-1246

[112] Dias ALB, Arroio Sergio CS, Santos P, Barbero GF, Rezende CA, Martínez J. Ultrasound-assisted extraction of bioactive compounds from dedo de moça pepper (*Capsicum baccatum* L.): Effects on the vegetable matrix and mathematical modeling. Journal of Food Engineering. 2017;**198**: 36-44

[113] Uquiche E, del Valle JM, Ortiz J. Supercritical carbon dioxide extraction of red pepper (*Capsicum annuum* L.) oleoresin. Journal of Food Engineering. 2004;**65**:55-66

[114] Sricharoen P, Lamaiphan N, Patthawaro P, Limchoowong N, Techawongstien S, Chanthai S. Phytochemicals in capsicum oleoresin from different varieties of hot chilli peppers with their antidiabetic and antioxidant activities due to some phenolic compounds. Ultrasonics Sonochemistry. 2017;**38**:629-639

[115] Cacace JE, Mazza G. Mass transfer process during extraction of phenolic compounds from milled berries. Journal of Food Engineering. 2003;**59**:379-389

[116] Fernández-Ronco MP, Gracia I, de Lucas A, Rodríguez JF. Extraction of *Capsicum annuum* oleoresin by maceration and ultrasound-assisted extraction: Influence of parameters and process modeling. Journal of Food Process Engineering. 2013;**36**:343-352

[117] Yilmaz E, Sevgi Arsunar E, Aydeniz B, Güneşer O. Cold pressed capia pepperseed (*Capsicum annuum* L.) oils: Composition, aroma, and sensory properties. European Journal of Lipid Science and Technology. 2015;**33**(8): 926-932

[118] Melgar-Lalanne G, Hernández-Álvarez AJ, Jiménez-Fernández M, Azuara E. Oleoresins from *Capsicum* spp.: Extraction methods and bioactivity. Food and Bioprocess Technology. 2017;**10**:51-76

[119] Kalil SJ, Moraes CC, Sala L, Burkert CAV. Bioproduct extraction from microbial cells by conventional and nonconventional techniques. In: Food Bioconversion. Vol. 2. London, UK: Elsevier Ltd; 2017. pp. 179-206

[120] Ventura SPM, Nobre BP, Ertekin F, Hayes M, Garciá-Vaquero M, Vieira F, et al. Extraction of value-added compounds from microalgae. In: Microalgae-Based Biofuels and Bioproducts: From Feedstock Cultivation to End-Products. Kidlington, UK: Elsevier Ltd; 2017. pp. 461-483

[121] Kim S, Chojnacka K. In: Kim S, Chojnacka K, editors. Marine Algae Extracts. Weinheim: Wiley-VCH; 2015. ISBN: 9783527333271

[122] Kapoore R, Butler T, Pandhal J, Vaidyanathan S. Microwave-assisted extraction for microalgae: From biofuels to biorefinery. Biology (Basel). 2018;**7**:18

[123] Lavilla I, Bendicho C. Fundamentals of ultrasound-assisted extraction. In: Water Extraction of Bioactive Compounds: From Plants to Drug Development. Kidlington, UK: Elsevier Ltd; 2017. pp. 291-316

[124] Gerde JA, Montalbo-Lomboy M, Yao L, Grewell D, Wang T. Evaluation of microalgae cell disruption by ultrasonic treatment. Bioresource Technology. 2012;**125**:175-181

[125] Baby KC, Ranganathan TV. Enzyme-assisted extraction of bioingredients. Chemical Weekly. 2013; **1**:213-224

[126] Alam MA, Sarker MZI, Ghafoor K, Happy RA, Ferdosh S. Bioactive compounds and extraction techniques. In: Recovering Bioactive Compounds from Agricultural Wastes. Chennai, India: John Wiley & Sons Ltd; 2017. pp. 33-53

[127] Nadar SS, Rao P, Rathod VK. Enzyme assisted extraction of biomolecules as an approach to novel extraction technology: A review. Food Research International. 2018;**108**: 309-330

[128] Rastogi NK, Raghavarao KSMS, Balasubramaniam VM, Niranjan K, Knorr D. Opportunities and challenges in high pressure processing of foods. Critical Reviews in Food Science and Nutrition. 2007;**47**:69-112

[129] Daher D, Le Gourrierec S, Pérez-Lamela C. Effect of high pressure processing on the microbial inactivation in fruit preparations and other vegetable based beverages. Agriculture. 2017;**7**(9): 1-18

[130] Tadapaneni RK, Daryaei H, Krishnamurthy K, Edirisinghe I, Burton-Freeman BM. High-pressure processing of berry and other fruit products: Implications for bioactive compounds and food safety. Journal of Agricultural and Food Chemistry. 2014; **62**:3877-3885

[131] Huang HW, Hsu CP, Yang BB, Wang CY. Advances in the extraction of natural ingredients by high pressure extraction technology. Trends in Food Science and Technology. 2013;**33**:54-62

[132] Torres JA, Velazquez G. Commercial opportunities and research challenges in the high pressure processing of foods. Journal of Food Engineering. 2005;**67**:95-112

[133] Dixon C, Wilken LR. Green microalgae biomolecule separations and recovery. Bioresources and Bioprocessing. 2018;**5**(14):1-24

[134] Poojary MM, Barba FJ, Aliakbarian B, Donsì F, Pataro G, Dias DA, et al. Innovative alternative technologies to extract carotenoids from microalgae and seaweeds. Marine Drugs. 2016;**14**(11):1-34

[135] Jubeau S, Marchal L, Pruvost J, Jaouen P, Legrand J, Fleurence J. High pressure disruption: A two-step treatment for selective extraction of intracellular components from the microalga *Porphyridium cruentum*. Journal of Applied Phycology. 2013;**25**:983-989

[136] Işlek C, Altuner EM, Alpas H. The effect of high hydrostatic pressure on the physiological and biochemical properties of pepper (*Capsicum annuum* L.) seedlings. High Pressure Research. 2015;**35**:396-404

[137] Hernández-Carrión M, Hernando I, Quiles A. High hydrostatic pressure treatment as an alternative to pasteurization to maintain bioactive compound content and texture in red sweet pepper. Innovative Food Science and Emerging Technologies. 2014;**26**:76-85

[138] Pereira CG, Meireles MAA. Supercritical fluid extraction of bioactive compounds: Fundamentals, applications and economic perspectives. Food and Bioprocess Technology. 2010; **3**:340-372

[139] Ehlers D, Czech E, Quirin KW, Weber R. Distribution of aflatoxins between extract and extraction residue of paprika using supercritical carbon dioxide. Phytochemical Analysis. 2006; **17**:114-120

[140] Škerget M, Knez Ž. Solubility of binary solid mixture β-carotene-capsaicin in dense CO_2. Journal of Agricultural and Food Chemistry. 1997;**45**:2066-2069

[141] Ricci A, Parpinello GP, Versari A. Recent advances and applications of pulsed electric fields (PEF) to improve polyphenol extraction and color release during red winemaking. Beverages. 2018;**4**(1):1-12

[142] Joannes C, Sipaut CS, Dayou J, Yasir SM, Mansa RF. The potential of using pulsed electric field (PEF) technology as the cell disruption method to extract lipid from microalgae for biodiesel production. International Journal of Renewable Energy Resources. 2015;**5**:598-621

[143] Ade-Omowaye BIO, Rastogi NK, Angersbach A, Knorr D. Combined effects of pulsed electric field pre-treatment and partial osmotic dehydration on air drying behaviour of red bell pepper. Journal of Food Engineering. 2003;**60**:89-98

[144] Won YC, Min SC, Lee DU. Accelerated drying and improved color properties of red pepper by pretreatment of pulsed electric fields. Drying Technology. 2015;**117**:1016-1026

[145] Ade-Omowaye BIO, Angersbach A, Taiwo KA, Knorr D. The use of pulsed electric fields in producing juice from paprika (*Capsicum annuum* L.). Journal of Food Processing and Perservation. 2001;**25**:353-365

[146] Campos-Vega R, Oomah BD, Vergara-Castaneda HA. Vegetable byproducts. In: Food Wastes and By-products: Nutraceutical and Health Potential. Pondicherry, India: John Wiley & Sons; 2019. pp. 254-255

11

Diversity and Potency of *Capsicum* spp. Grown in Indonesia

Cristofora Hanny Wijaya, Manik Harda and Bunga Rana

Abstract

Capsicum spp., popularly known as chili pepper, is abundantly cultivated in Indonesia. Chili pepper has deeply integrated into Indonesian culture, even turned into inseparable ingredients in the local diets. Adapted to the Indonesian environment and timely selected according to the local palate, vast variation of appearance, color, taste, aroma, and pungency of chili pepper transformed into a variety of cuisines and medicinal purposes. This chapter covers a selection of chili peppers in Indonesia, introduced as "commonly available in the market," and some other variants, which are less popular; however, due to their unique flavors, they will be interesting to be intro-duced and addressed as "exotic *Capsicums*." This chapter will describe their botanical information, distribution area, and nutritional and flavor aspects, including their bioactive compounds. Furthermore, the utilization of genus *Capsicum* in Indonesia, from the food industry, pharmaceutical, and ornamental plants, will be described.

Keywords: bioactivity, *Capsicum* spp., chili pepper, flavor, Indonesia

1. Introduction

Portuguese brought *Capsicums* to the eastern part of Indonesia, Maluku, in the seventeenth century and then widely distributed to other parts of the archipelago. It is assumed that the four species (*C. annuum*, *C. frutescens*, *C. chinense*, and *C. pube-scens*) were introduced to Indonesia before World War II [1]. During that time, the Javanese called the plant "godong sabrang," which means a plant from a foreign coun-try [2]. Later, in each region of Indonesia, *Capsicums*, especially chili or hot peppers might have different names, like "cengek," "lado," "lada," "cabe," "cabi," "lombok," "ricagufu," "serbeh ulom," "serbeh," and many others. However, officially, pungent members of *Capsicums* or hot chili peppers in Indonesia are called "cabai," while the sweet and nonpungent ones are called "paprika." From then on, in this text, we use the term "cabai" for hot peppers or chili peppers and "paprika" for sweet peppers.

The *Capsicums* currently cultivated in Indonesia are quite different from those first carried by the Europeans since they have been crossbred and undergone genetic changes due to adaptation to local environments and natural and human selection. The seeds of *Capsicums* cultivated in Indonesia are also distributed to neighboring countries and even to Macau, China [3, 4]. During that time, local inhabitants of Java Island used *Capsicum* fruits in their cooking, while the leaves were used for topical ointment.

Indonesia also included *Capsicums* as part of their traditional ceremony and ritual. In some ceremonies in Java, sometimes, they put *Capsicums* on top of *tumpeng*, a traditional dish consisting of cone-shaped yellow-colored rice complemented with other side dishes and vegetables. Examples of traditional Indonesian dishes using *Capsicums* are *sambal terasi* (Java), *balado Padang* (West Sumatra), *rica-rica Manado* (North Sumatra), and *lawar* (Bali). *Lawar* from Bali was involved in wedding, death, and other religious ceremonies [2].

Capsicum has become an essential commodity in Indonesia, where its price might influence political and socioeconomic conditions. Similar to what happened in South Korea in 1978–1979, where the people's unrest was subjected to the government due to a shortage of *Capsicums* [5], in Indonesia, the price of *cabai* could be one of the parameters to measure the government performance in establishing food security. For the majority of Indonesians, regardless of social class or economic status, *Capsicum* has become an inseparable part of their daily consumption and may serve as a source of vitamin C, E, and folate, as well as carotenoids, phenolic compounds, flavonoids, and capsaicinoids [6, 7]. Even with the decrease in household incomes or an increase in household expenses, consumers still tend to buy *cabai* [8, 9]. Like India, Thailand, and South Korea [5, 10], the consumption of *Capsicums* in Indonesia was also high. Among other vegetable plantations, *cabai* occupied 416,3981 ha of cultivation area in 2019 [11], the second largest in the country after cabbage.

Cabai prices could influence the price of other food products, affecting the economy, consumption, and buying habits of the consumers. Hence, the fluctuation of *cabai* prices could trigger inflation. For example, in the rainy season, there is a possibility of decrease in *cabai* production due to the crop failure. The unmet high demand, causing a double increase in price for *cabai*, resulted in inflation. This phenomenon has already occurred for several years. *Cabai* contributed 0.09% of 0.28% Indonesian inflation level in the early rainy season in October 2018, and *cabai* was also reported as the leading cause of Indonesian economic inflation in 2006 based on data from Statistics Indonesia [13].

2. Types of *Capsicum* species in Indonesia

From the genus *Capsicum*, there are 27 species members, including five domesticated species (*C. annuum, C. frutescens*, *C. chinense*, *C. pubescens*, and *C. baccatum*) around the globe [14]. *Capsicum* species commercially cultivated in Indonesia are *C. annuum* and *C. frutescens* as seen in **Figure 1**. The other species like *C. chinense* and *C. pubescens* can only be found in several districts and usually cultivated by

Figure 1.
Capsicums *commonly available in Indonesian market.*

hobbyists [4]. From the distribution of *C. chinense* in Indonesia, though it is very limited in Southeast Asia, it is assumed that "Indonesia had more potential and genetic resources to breed species of *Capsicum* than other countries in Southeast Asia" [1].

In Indonesia, there are *Capsicum* hybrid lines produced by seed agents, like "cabai sultan," "hot beauty," "megahot," "Carvi Agrihorti," and "cabai trisula." *Capsicum* hybrid lines were most farmer's favorites, due to their high productiv-ity beside their relatively high prices. However, as progenitors, they might not be good candidates due to their variability in the progenies. Meanwhile, the local and open-pollinated cultivars might not produce as much yield as hybrid lines, but they are generally more resistant to pests and disease, cheaper, and fit to use as progeni-tors [15]. The cultivated variety is selected based on market demand (color, aroma, taste, pungency, appearance, and size), high productivity, resistance to pests and disease, and adaptability to the local environment. Meanwhile, the consideration in choosing seeds is the availability of the seed's certificate, moisture content, seed purity, germination, absence of contaminants, and seed's health.

2.1 Commonly available in the market

*2.1.1 Cabai keriting (*Capsicum annuum *L. var. Longum)*

Cabai keriting is a highly marketable commodity. The fruits are thin-skinned, long (around 12 cm), slender, curly, or wavy in appearance with pointed tips as shown in **Figure 2** [16]. There are many seeds inside the fruit. The plants usually have a productive age up to 6 months in relatively high humidity with a temperature around 16–32°C, soil pH around 5–7 [17], and relatively resistant to plant diseases and can be harvested all year round [4]. On average, each plant can produce up to 0.75–1 kg fruit/harvesting period, i.e., 6 months. The fruits are harvested both in the immature stage (green) and the mature stage (red) [18].

This cultivar of *Capsicum* is one of the superior and popular cultivars in Indonesia due to the plants' productivity and relatively high price, and their thin skin and lower moisture than *cabai merah besar* make them not so prone to rot and have longer shelf-life compared to *cabai merah besar* [4, 17]. Some known varieties are *cabai keriting Solok*, *cabai keriting Cirebon*, *cabai keriting Irian*, *cabai keriting ungu*, *cabai keriting Lampung*, *cabai keriting Padang*, *cabai keriting Bengkulu*, *cabai*

Figure 2.
Cabai keriting *(*C. annuum *L. var. Longum).*

keriting Lembang, *cabai keriting Pangalengan*, and *cabai keriting Medan* [19, 20]. In 2011, there were 86 cultivars of *cabai keriting* in total [21]. From the plant morphology, there were 13 parameters to consider in choosing cultivars, i.e., leaf shape, leaf tip, the color of the anther, the color of pistil, fruit surface, the color of young fruit, plant height, dichotomous height, canopy width, fruit length, fruit diameter, and fruit weight [15]. Examples of popular *cabai keriting* seeds cultivated in Indonesia are Taro, Lado, Bagayo, Tanamo, Jago, New Taro, and Laris. A picture of *Cabai keriting* grown in Indonesia is shown in **Figure 2**.

Distribution areas. *Cabai keriting* is usually cultivated in West Java and Sumatra Island (Bengkulu, Lampung, West Sumatra, and North Sumatra) [16, 19].

Flavor characteristics. The taste of *cabai keriting* is relatively hotter than *cabai merah besar* (*C. annuum*) but not as hot as *cabai rawit* (*C. frutescens*). There were several varieties of *cabai keriting* in Indonesia, e.g., *cabai keriting ungu*, *cabai keriting Lampung*, *cabai keriting Padang*, *cabai keriting Bengkulu*, *cabai keriting Lembang*, *cabai keriting Pangalengan*, and *cabai keriting Medan*, according to the order of their pungency from low to high with Scoville heat unit (SHU) varied from 30,000 to 170,000 [19]. Since the flavor components change during maturation [22], it would be important to notice that the green and red maturing stage of cabai keriting might alter the character of the dishes. For example, *ayam cabe hijau* was a traditional cuisine that is uniquely characterized by the use of immature stage (green) of *cabai keriting* as the main seasoning for cooked chicken. Meanwhile, red *cabai keriting* is usually used in cooking for its attractive red color and relatively hot pungency.

Bioactivity and health benefits. *Cabai keriting* was believed to have antimicrobial effects; thus, the inhibitory effect of the oleoresin of *cabai keriting* against *Staphylococcus aureus*, *Staphylococcus epidermidis*, *Escherichia coli*, and *Pseudomonas aeruginosa* had been reported [23]. The oleoresin contained tetramethyl hexadecenol, tetramethyl hexadecene, tetramethyl hexadecapentaene, ethyl decahexanoate, ethyl linoleate, methyl linoleate, palmitoyl chloride, hexanedioic acid, octadecadi-enoic acid, ethyl linoleolate, butyl isopropylphenyl oxalate, and capsaicin.

*2.1.2 Cabai merah besar (*Capsicum annuum *L.) and (*Capsicum annuum *L. var. Tanjung)*

The fruits of *cabai merah besar* are quite long, big, and bulky with a smooth surface and thick skin [12]. Some fruits have sharp points, and the others have round points [24]. On average, each plant can produce up to 1–1.2 kg fruit/harvesting period [18], but this number depends on the varieties, agricultural practices, and plants' environments. The farmers harvest both immature fruits (green) and mature fruits (red) of *cabai merah besar*.

Some known varieties of *cabai merah besar* are Ciko, Lingga, Tanjung, Kencana, Tit Super, Paris, and Jati Laba. In 2011, there were 87 cultivars of cabai merah besar in total [21]. Examples of popular *cabai merah besar* seeds cultivated in Indonesia are Profit, Arimbi, Horison, Imperial 10, Jet Set, Imola, Gadewa, Gada, Elegance, and Gada MK. *Cabai merah besar* is shown in **Figure 3**.

Distribution areas. *Cabai merah besar* is cultivated in almost every province in Indonesia especially in West Java, Central Java, North Sumatra, East Java, West Sumatra, Aceh, Lampung, South Sumatra, South Sulawesi, and Bengkulu [25].

Flavor characteristics. There were several varieties of *cabai merah besar* in Indonesia, e.g., Tit Super, Paris, and Jati Laba, with pungency around 12,500 SHU [19]. *Cabai merah besar* is the least hot compared to *cabai keriting* and *cabai rawit* (*C. frutescens*). The popular cultivar of *C. annuum* was less hot compared to *C. frutescens* [26]. Since their pungency is considered low, *cabai merah besar* is usually used due to its attractive color, texture, and aroma, not because of their pungency.

Figure 3.
Cabe merah besar *(*C. annuum *L.) in the market.*

The fruits have a crunchy texture and higher moisture content compared to *cabai keriting*. However, the same with *cabai keriting*, both green and red maturing stages of *cabai merah besar* are also used, but the application is different depending on the intended dishes. The green stage of *cabai merah besar* is usually boiled, stir-fried, or fried before consuming, to remove unwanted aroma.

Nutrition component. *Cabai merah besar* contains moisture (90%), energy (32 kcal), protein (0.5 g), fat (0.3 g), carbohydrate (7.8 g), fiber (1.6 g), ash (0.5 g), calcium (29.0 mg), phosphorus (45 mg), iron (0.5 mg), vitamin A (470 IU), vitamin C (181 mg), thiamin (0.05 mg), riboflavin (0.06 mg), and niacin (0.9 mg) per 100 g red fruits [27–29], while the green fruits contain energy (23 kcal), protein (0.7 g), fat (0.3 g), carbohydrate (5.2 g), calcium (14 mg), phosphorus (23 mg), iron (0.5 mg), vitamin A (260 IU), and vitamin C (84 mg) per 100 g fruits [24].

Bioactivity and health benefits. It is reported that the ethanolic extract of *cabai merah* with a dosage of 200, 400, and 600 mg/kg was effective in lowering blood sugar levels in male white mice hyperglycemia [30]. The most effective duration in lowering blood glucose levels was 21 days. The possibility of luteolin extracted from *C. annuum* L. as an alternative to treat autism spectrum disorder (ASD) using the reverse docking technique has been explored; the result showed that luteolin had a potency to work as an MMP-3 antagonist. MMP-3 (matrix metalloproteinase) was suspected of having a role in the neuropathology of ASD [31]A study about *cabai merah* effects towards hypercholesterolemia showed that 200 mg/kg ethanolic extract of *cabai merah* had a cholesterol-lowering effect on hypercholesterolemia mice with the best administration duration in lowering blood cholesterol of male albino mice of 14 days [32]. The conversion of this dose to humans with the average body weight of 70 kg results in as much as 173.15 g of *cabai merah* needed to get the expected effect. This bioactivity might be attributed to sev-eral compounds such as capsaicin, vitamin C, saponin, and flavonoid compounds. It is reported that dietary capsaicin improved endothelium-dependent vasorelaxation and prevented hypertension [33].

The inhibitory effect of oleoresin of *cabai merah besar* against *Staphylococcus aureus*, *Staphylococcus epidermidis*, *Escherichia coli*, *and Pseudomonas aeruginosa* was also reported. Oleoresin contained tetramethyl hexadecenol, tetramethyl hexadecene, tetramethyl hexadecapentaene, ethyl hexadecanoate, linoleic acid, methyl lino-leate, azacylotridecanone, hexanedioic acid, octadecadienoic acid, ethyl linoleolate, octadecenyl acetate, and capsaicin [23].

2.1.3 Paprika (Capsicum annuum *var. grossum)*

The fruit color and shape vary as shown in **Figure 4**, but mostly they have bell-shaped, long with pointed tips or almost square with hollow space inside, and thick skin (around 0.5 cm). The fruit is relatively big compared to *cabai merah besar* [12, 34]. The seeds are clustered together and attached to the placenta [35]. On average, each plant can produce around 3.5–3.7 kg fruit/harvesting period, i.e., 12 months [18]. There are many paprika seed varieties cultivated in Indonesia, for example, Jumbo seed, Takii ace, Green Horn, New Ace, Wonder Bell, Top Star, Blue Star, Uranus, Big Star, Queen Star, Beauty Bell, Sunny Star, Lucky Star, Colombo, Mayata, Golden California Wonder, Skipper, Melody, Fortuna, Yellow Star, and Virgo [35].

Distribution areas. Paprika is usually cultivated in North Sumatra, Riau, West Java, East Java, Bali, West Nusa Tenggara, East Nusa Tenggara, Central Sulawesi, Southeast Sulawesi, Maluku, and West Papua [25].

Flavor characteristics. The fruit tastes mildly sweet with a crunchy texture. The fruit aroma has a pungent impression but does not taste hot at all [34]. The fruit slices are usually cooked with meat or as pizza toppings.

Nutrition components. The fruit contains protein 0.9%, fat 0.3%, carbohydrate 4.4%, calcium 7 mg/100 g, phosphorus 22 mg/100 g, iron 0.4 mg/100 g, vitamin A 22 IU/100 mg, vitamin B1 540 mg/100 g, vitamin B2 0.02 mg/100 g, vitamin C, 160 mg, and niacin 0.4 mg/100 g [35].

Bioactivity and health benefits. As an herbal medicine, consuming paprika is suggested to control cholesterol and triglycerides and prevent atherosclerosis, hypertension, flu, sinusitis, and digestion problem [36]. One study found that the ethano-lic extract of paprika exhibited antimicrobial activity towards pathogen *Enterococcus faecalis* [37]. Another study reported that *Capsicum annuum* var. grossum could be

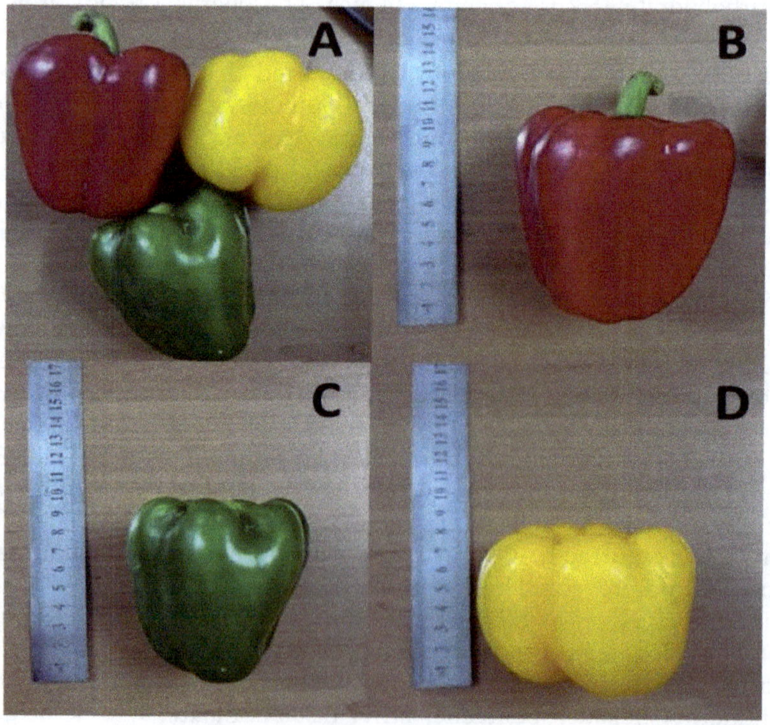

Figure 4.
(A) Some paprikas commonly available in Indonesian market, (B) red paprika (C) green paprika, (D) yellow paprika.

an alternative intervention for Alzheimer's disease (AD) by inhibiting β-secretase activity and β-Amyloid$_{1–40}$ aggregation [38]. It is stated that phenolic extracts from the fruits could counteract the initial aggregation of Aβ$_{1–40}$, as well as prevent further aggregation of preformed fibrils. The deposition of amyloid protein as senile plaques produced by the sequential cleavage of the amyloid precursor protein by secretases was the key signature of Alzheimer's disease. These inhibitory activities were attrib-uted to the presence of phenolic constituents in the fruits [38].

*2.1.4 Cabai rawit (*Capsicum frutescens*)*

As it is shown in **Figure 5**, *cabe rawit* fruits are small, around 3.7–5.3 cm long, and the seeds are usually light brownish [12]. On average, each plant can produce a 0.5–0.6 kg fruit/harvesting period [18]. There are several cultivars of *cabai rawit* in Indonesia. A bit bulky with round tips is *cabai ceplik*. Immature *cabai ceplik* in Indonesia turned color from bright green to dark red during maturation. *Cabai rawit* with relatively higher pungency than *cabai ceplik* is *cabai jemprit*. The fruit is small and short and has a sharp point. The fruit turns color from dark green to red during maturation. *Cabai putih/cabai rawit domba* looks similar to *cabai jemprit*, but the color is yellowish and turns bright reddish orange during maturation [24]. There are also other cultivars called *cabai kathur* because the fruit's tip points to the sky and *cabai burung* or bird eye pepper with small fruits. There are several seed varieties cultivated in Indonesia, e.g., Pelita 8, Sonar, Bara, Taruna, Dewata, Nirmala, Lentera, Raga, Bhaskara, Maruti, Cakra Putih, and Cakra Hijau.

Distribution areas. *Cabai rawit* is cultivated in almost every province in Indonesia, especially in West Java, Central Java, East Java, Aceh, West Nusa Tenggara, and Bali [25].

Flavor characteristics. The taste of *cabai keriting* is relatively hotter than *cabai merah besar* (*C. annuum*) but not as hot as *cabai rawit* (*C. frutescens*). There were several varieties of *cabai rawit* in Indonesia, e.g., *rawit putih*, *cabai rawit hijau*, *cabai rawit Lampung*, and *cabai rawit Kalimantan*, according to the order of their pun-gency from low to high with 82,500–480,000 SHU [19], the major organic acids of *C. frutescens* were malic, citric, and ascorbic acid. In contrast to *C. annuum* in which citric acid was the primary organic acid, in *C. frutescens*, the dominant organic acid

Figure 5.
*Cabai rawit (*C. frutescens*).*

was malic acid [26]. The volatile compounds of *C. frutescens* consisted of aliphatic esters and methyl salicylate imparting fruity and fresh impressions. However, as the fruit turned to red, the flavor changed towards more pungent and citrus-like aroma. The pungency also increased as the fruit matured [39], but the pungent compound might decrease at later stages of maturation [40]. The immature fruit, having less pungency than the mature ones, is used to be consumed raw with vegetable fritter snack, fried tofu, and tempeh or put into rissoles.

Nutrition components. *Cabai rawit* contains energy (103 kcal), protein (4.7 g), fat (2.4 g), carbohydrate (19.9 g), fiber (1.6 g), ash (0.5 g), calcium (45.0 mg), phosphorus (85 mg), vitamin A (11,050 IU), and vitamin C (70 mg) per 100 g fruits [24].

Bioactivity and health benefits. *Cabai rawit* contains phytochemicals such as flavonoids, capsaicinoids, and other phenolic compounds [41]. Capsaicinoids have been reported to have various health benefits such as antioxidant, anticar-cinogenic, anti-inflammatory, and relieving neuropathic pain [42–45]. The locals believe that *cabai rawit* is good for the eyes, can cure cancer, alleviate joint pain, improve blood circulation, and stimulate appetite [46]. The potency of *C. frutescens* extract for anti-obesity in 3 T3-L1 cells, due to its higher content of phenolic compounds and antioxidant activity compared to *C. annuum*, was explored [39].

2.2 Exotic *Capsicum*

*2.2.1 Cabai gendot (*Capsicum pubescens*)*

Before 1916, this *Capsicum* was introduced to Indonesia, and the locals called this *cabai* as "cabai gendot," "cabai bendot," "cabai gondol," "cabai gombol," "cabai Bandung," or "cabai Dieng" [47]. The fruit shape varies from round to globular, 3.5–4 cm long, and 2.5–4 cm in diameter with thick flesh [40]. The fruit color changes from green (immature) to red (mature). The seeds are relatively small and few. Some people might confuse this *cabai* as *C. chinense*. The picture of *cabai gendot* can be seen in **Figure 6**.

Distribution areas. *Cabai bendot* is usually cultivated in Bandung (West Java), Dieng (Central Java), Kota Batu, and Pasuruan (East Java). This *cabai* is rarely found in the market [16, 47].

Flavor characteristics. The flavor of *cabai gendot* differs from *C. annuum*. This *Capsicum* cultivar is relatively juicier and hotter than *cabai merah besar*, around 100,000–350,000 SHU. The fruit has a fruity and floral aroma [34]. Due to their hotness, some people rarely eat fruit raw. People used to put both mature and immature stages of *cabai gendot* in *sambal* and mix them in stir-fry vegetables (*oseng-oseng*), or other plant-based dishes (e.g., *sayur tahu*, *sayur asem*). The fruits also have a strong and unique pungent aroma [47].

Figure 6.
*Cabai gendot (*C. pubescens*) (A) red cabai gendot (B) green cabai gendot.*

*2.2.2 Cabai katokkon or cabai kotokkon (*C. chinense*) or (*Capsicum annuum *L. var. sinensis)*

The name *cabai katokkon* was derived from the word "katokkon," which means "big fruit." This *cabai* is classified as *C. chinense* [1] or *C. annuum* L. [48]. The fruits are shaped like paprika but smaller (around 3–4 cm in length and 2–3.5 cm in diameter), round, and bulky as shown in **Figure 7**. The immature fruit is purplish green and turned bright red when mature. This plant can grow well on 200 m above sea level with adequate sunlight and watering. The plant can be harvested 3 months after seed planting, and on average, the plant can produce 0.8–1.2 kg fruits/harvest-ing period, i.e., 8–10 months [49].

Distribution areas. *Cabai katokko*n is cultivated in Rantepao, Tana Toraja (South Sulawesi) [1].

Flavor characteristics. The fruits have a unique floral, fruity and strong pungent aroma, and higher pungency than commonly consumed *Capsicums* [1]. By calculating the concentration of capsaicin (without taking account of dihydro-capsaicin), the pungency level of cabai kotokkon dried powders was reported to be around 42.914 SHU (immature stage) and 48.778 SHU (mature stage) [50]. Other sources reported the pungency level of 400,000–691,000 SHU [48].

Bioactivity and health benefits. Besides containing vitamin C and provitamin A that have antioxidant capacity, in Tana Toraja, South Sulawesi, people believe that the fruits could increase appetite, prevent aging, reduce stress, help to alleviate joint pains, lower cholesterol, improve blood circulation, prevent stroke, and alleviate wet cough, nasal congestion, and migraine [49].

*2.2.3 Cabai domba (*Capsicum frutescens*)*

The fruit shape is small but a little bit bigger than common *cabai rawit*. It has a round tip as shown in **Figure 8**. The color of the fruit varies from greenish white (immature stage), pale green, and orange [51]. Some people put them in the same category as *cabai rawit*; *thus*, it is called "cabai rawit domba" or "cabai putih," and some other people confused cabai domba as *Capsicum annuum* L. var. abbreviata Fingerhuth.

Distribution areas. *Cabai domba* is cultivated in Bandung, West Java [51].

Flavor characteristics. The pungency of the fruit is around 50,000–100,000 SHU [49]. Due to their relatively hot taste, the fruit is rarely consumed raw; instead,

Figure 7.
*Cabai katokkon (*C. chinense*) growth in the garden.*

Figure 8.
*Cabai domba (*C. frutescens*).*

Figure 9.
*Cabai hiyung (*C. frutescens*) [54].*

people mixed it in their cooking or put in *sambal*. They have a strong pungent smell, especially when exposed to heat.

2.2.4 *Cabai hiyung (*Capsicum frutescens *L.)*

Cabai hiyung has an appearance similar to *cabai rawit*, around 2–3 cm in length, and relatively thin skin with the weight of each fruit is around 0.7 g [52]. The plant can be harvested at the age of around 3–4 months with harvesting period of 6–7 months. The fruits are harvested when they turn brownish yellow or red [53]. The picture of *cabai hiyung* [54] can be seen in **Figure 9**.

Distribution areas. *Cabai hiyung* is cultivated in Hiyung village, South Kalimantan [53].

Flavor characteristics. The fruit was reported as having many times higher pungency than the common cultivar of *C. frutescens* in Indonesia; therefore it was named as the hottest *C. frutescens* in Indonesia [55]. Though we could not find the level of pungency in SHU, the fruits reported having 802 ppm of capsaicin [56]. **Nutrition components.** The fruit was reported containing vitamin A (11.89 IU/100 g), 82–92 mg/100 g vitamin C, and 1.2–1.5% protein [53, 56].

2.2.5 *Cabai ceremai (*Capsicum chinense*)*

This *cabai* is a habanero-type pepper (*C. chinense*), in which the colors vary from green, yellow, bright orange, and orange. Different names have been attributed to this fruit, such as "cabai ceremai," "cabai cermai," "cabai kancing," "cabai tomat," "cabai belimbing," "baby chili Haba," and "cabai tawau." The fruits are generally clustered into three to five pieces in each branch. The fruit shape varies from round shape with a papillae tip to bell and cone and long shape with fruit length up to 12 cm. The fruit skin wrinkles, but sometimes it also has smooth skin. The colors of the fruits vary from green, red, orange, pink, yellow, or brown. The seeds have a pale color [1, 4].

Distribution areas. The distribution of *cabai ceremai* in Southeast and East Asia was still limited. *Cabai ceremai* can be found in Jakarta, Bogor, Sukabumi (West Java), Jepara, Tawangmangu (Central Java), Malang (East Java), Bangka Island, Bali, Putu Sibau (West Kalimantan), and Tarakan (East Kalimantan) [1, 4].

Flavor characteristics. The fruit was reported to have high pungency [1, 4], but we could not find the data on the level of pungency in SHU. *Cabai ceremai* was not popular in the market since consumers preferred to buy and consume *cabai* that have elongated shape [57].

3. *Cabai* utilization and applied products in Indonesia

Since its first coming to Indonesia, *cabai* has been widely utilized in the food and culinary industry. Besides that, *cabai* in Indonesia was also utilized in the pharmacy as transdermal medicine (plaster, oil, balm, and cream) and as ornamental plants due to its attractive and unique fruit colors.

3.1 *Cabai* in Indonesia food and culinary culture industry

Cabai is considered as an important ingredient since it is inseparable from Indonesian food culture as is shown in some of Indonesia dishes in **Figure 10**. During a meal, if *cabai* is not included in the dish seasoning, it will be served separately as "*sambal*," crushed *cabai* mixed with various other ingredients and commonly eaten alongside the main course, or the *cabai* will be served to be eaten raw. Some people even consider something missing in their meal, and it would not be "complete" if their food is not spicy or there is no *sambal* to be eaten with the main course. Therefore, their satisfaction with the food will decrease. Over time, the way of people consuming *cabai* was developed, and *cabai* was transformed into processed food products, such as sauce and powder, as the complementary dish of the main course (**Figure 11**).

The utilization of *cabai* in local cuisines may vary from vegetable-based food such as stir-fry vegetables or as a sauce (or like salad dressings) together with peanut in mixed vegetables, such as *gado-gado*, *pecel*, and *karedok*. *Cabai* is also mixed in meat-, chicken-, and seafood-based cuisines, such as *rendang* (Indonesian famous pungent cuisine), *gulai* (spicy curry), *asam padeh* (spice and acid fish curry), and other dishes. "Balado" is an Indonesian dish that combines any type of food ingredi-ents with *cabai merah*, for example, egg *balado*, fish *balado*, chicken *balado*, *dendeng* (dried meat) *balado*, eggplant *balado*, and shrimp *balado*.

3.1.1 Sambal: An Indonesian traditional crushed cabai

Sambal is an Indonesian famous and essential complementary cuisine made from *cabai* as the main ingredient. It is a crushed or grounded *cabai* with or without any other ingredients. Generally, *sambal* is known to stimulate appetite or to increase food palatability [8]. Before *Capsicum* came to Indonesia, *sambal* was originally made from ginger. Then people started to prepare *sambal* from pepper (*merica*) and *cabai Jawa* (*Piper retrofractum*) [58] Since *Capsicum* spp. was introduced to Indonesia and cultivated, *sambal* was dominantly made from *Capsicum* spp.

Figure 10.
Cabai *as an inseparable ingredient in Indonesian cuisines. (A)* Cabai *slices are mixed as ingredient in daily cuisine, (B)* Sambal *and* cabai sauce *in meat balls cuisine, (C)* Cumi balado *(calamari* balado*) as an example of* balado *dish, (D)* Rendang, *Indonesian famous spicy cuisine with* cabai *as one of essential ingredient.*

Figure 11.
Common cabai processing product in Indonesia. (A) Sambal terasi, (B) cabai sauce, (C) cabai powder.

Almost every region in Indonesia has their own types of *sambal*. There were about 322 types of sambal in Indonesia, and they varied over the nation with 119 types of raw sambal, 138 types of cooked sambal, and other types of *sambal* [59]. *Sambal terasi*, as seen in **Figure 11(A)**, was made from *cabai merah*, *cabai rawit*, *terasi* (fermented shrimp), onion, garlic, salt, and sometimes tomato. *Sambal roa* from Manado, North Sulawesi, was made from *Capsicum* and smoked *roa* fish, which can be found abundantly in some particular areas. *Sambal tempoyak* from southern Sumatra was mainly made from *Capsicum* and *tempoyak*, a fermented durian. Other examples of *sambal* variants are mango *sambal*, *andaliman sambal*, *matah sambal*, various kinds of fish and seafood *sambal,* and sweet soy sauce *sambal*. The local culture and commodities available in the area influenced the ingredients included in *sambal*.

3.1.2 Cabai sauce

Cabai sauce or "saus cabai," as seen in **Figure 11(B)**, is described as a sauce made from *Capsicum* sp. as the main ingredient and processed with or without the addi-tion of other legitimate food additives [60]. *Saus cabai* is usually made from *cabai*, salt, onion, and garlic, with the addition of cornstarch as a thickening and binding agent [61]. The ingredients are sorted and washed, and then, *cabai* is steamed, mixed, and grounded with other ingredients. Cornstarch solution is added, and the mixture is cooked. Acidity regulators and preservative agents are added and mixed before *saus cabai* are packed and stored [61]. The flavor of *saus cabai* varies from sweet to spicy, but its pungency level is usually lower than *sambal*.

3.1.3 Cabai powder

Cabai powder, as seen in **Figure 11(C)**, is used as an ingredient to bring up spicy taste in savory processed food such as instant soup, instant noodles, and snacks. In 2012, a company developed a mixture of *cabai* powder with other ingredients such as onion, garlic, pepper, and also with particular fish and shrimp. The addition of fish and shrimp has enriched the umami taste of *cabai* powder products in Indonesia. "Tasty chili powder" with umami and spicy taste is developed as an innovation of *cabai* powder, and a new habit of eating *cabai* emerged in Indonesia with this dry powder as a condiment to be sprinkled on top of the food instead of dipping or mixing the food in *sambal*. The spiciness of tasty chili powder varies up to several levels from low to high. In the making of this powder, the ingredients were sorted, blanched, dried, grounded, and packed [61]. Tasty chili powder contained up to 16.8 ppm capsaicin [62].

3.2 Other rare and unique foods from *Capsicum* in Indonesia

3.2.1 Manisan cabai/sweetened capsicum

Manisan cabai is sugar-sweetened *Capsicum* slices, which is commonly eaten as a snack. It is usually made from thick-skinned red *Capsicums* such as *cabai merah besar* (*C. annuum* L.). The processing of sweetened *Capsicum* started from seed removal. The remaining skin is soaked in the salt solution to remove the pungent flavor. Then, *Capsicum* is soaked in $CaCO_3$ solution to harden the texture before being soaked in the warm sugar solution for 6–12 hours to absorb the sugar. The sweetened *Capsicum* may be dried and sprinkled with sugar. Then the resultant products are being packed and stored. Besides altering *Capsicum's* taste by masking the pungency and increasing the sweetness, which makes *Capsicum* likable by children, this process also lengthens the shelf life of *Capsicum* due to high sugar concentration. The example of sweetened *Capsicum* is shown in **Figure 12**.

3.2.2 Capsicum chocolate bar

Capsicum chocolate bars tasted like typical chocolate bars at early bites, but the pungent taste of *Capsicum* will appear after several bites. Although *cabai rawit* (*C. frutescens*) used in the chocolate bar has quite strong pungency level, it can be covered by the high content of chocolate (58–60%) [63]. An example of *Capsicum* chocolate bar can be seen in **Figure 13**.

Figure 12.
*Sugary sweetened Capsicum (*Manisan cabai*)* .

Figure 13.
Capsicum chocolate bars.

*3.2.3 Kopi cabai rawit/*Capsicum frutescens *coffee*

Kopi cabai rawit is a mixture of coffee, milk, ginger, and *cabai rawit* (*C. frutescens*). This beverage can be found in Bandung, West Java, Indonesia. This coffee has a unique taste from the bitterness of caffeine, pungency from ginger, and spiciness from *cabai rawit.* After drinking the coffee, people felt a warmth sensation towards the throat and the body [64].

3.3 Pharmacy

Indonesia is rich in spices and herbs, and the locals have been utilizing them as traditional medicine for years. This tradition has been developed, as the active ingredients from traditional medicine are processed into more convenient products with the same function and characteristics. Plasters/patches, cream, balm, and oil with pungent and mint sensations have been used daily for muscle pain, headache, stomachache, and reducing the pain of toothache. The applications of those products from *Capsicum* are verily nurtured in Indonesian culture.

Capsaicin, responsible for pungent taste in *Capsicum*, also plays a role in the hot sensation in the skin. This hot sensation is utilized in the pharmacy industry as

Figure 14.
Koyo cabai, *the* Capsicum *hot medicated plaster in Indonesia.*

a muscle pain reliever by adding plaster fabric, cream, or gel, together with other supporting ingredients. Besides that, capsaicin cream is useful for chronic soft tissue pain and chronic back pain. It gives beneficial effects as first-line therapy of osteoarthritis pain and decreases post-surgical neuropathic pain [65–67].

3.3.1 Koyo cabai or capsicum transdermal patch/capsicum hot medicated plaster

Koyo cabai or *Capsicum* Hot Medicated Plaster is one of medicated plasters in Indonesia. It is usually used for the treatment to relieve muscle pain or reduce the pain of toothache. *Koyo cabai*, at first imported from a Deutsch company, had been used in Indonesia since 1970. However, in 1983, a local company started to produce *koyo cabai* [68]. The plaster was made from high concentration*cabai* extract mixed with rubber as sticking agent and other ingredients. The sticky mixture then applied and pressed to the plaster fabric layer. The layer was cut and packed. *Koyo cabai* as a famous *Capsicum* Medicated Plaster in Indonesia can be seen in **Figure 14**. One Capsaicin Medicated Plaster may contain 0.37 mg/cm^2 of *Capsicum* extract [69].

3.3.2 Cabai cream and cabai balm

Similar to *koyo cabai*, *cabai* cream or *krim cabai* and *cabai* balm or *balsam cabai* in Indonesia are mainly utilized to relieve muscle pain and headache. It is a common thing that people in Indonesia keep muscle pain-relieving cream or balm as a first aid medicine in their homes or travel kits. *Cabai* cream may contain 1.66% *cabai* oleoresin; meanwhile, *cabai* balm may contain 0.56% *cabai* oleoresin.

3.4 Plant ornaments

Ideally, ornamental *cabai* has bonsai-like quality that is short in height (dwarf) and produces quite a number of fruits with various colors [70]. *Capsicum* species, especially *C. chinense* and *C. pubescens*, are attractive to be used as ornamental plants. *C. chinense* has various fruit shapes that look like a combination of tomato and starfruit and different bright colors during maturation, while *C. pubescens* has purplish fruits and is combined with erect flowers [71].

From the ornamental plants *C. annuum* L., the variety is called "Ayesha IPB" [70]. The plant relatively early blossoms in 13–16 days after planting and can be harvested in 65–70 days after planting. The average plant height is 25 cm with 46–53 cm of canopy width, and the leaves of canopy are nicely clustered. The fruits have round tips, and they would turn from yellow green (immature stage) to orange (intermediate) and orange red (mature stage). The other variety is called

Figure 15.
Some Capsicum ornamental plants in Indonesia (A) Cabai Ayesha *(B)* Cabai Syakira *(C)* Cabai Lembayung *[73].*

"Lembayung IPB" with a short height, purplish stem and stalk, and attractive fruit colors. The immature fruit color is purple and they would turn into dark orange [72]. The fruit tastes relatively hot. Other ornamental cultivars from *C. annuum* are "Namira," "Syakira," "Jelita," and "Ungara." Some types of *cabai* as ornamental plants [73] can be seen in **Figure 15**. Colorful *Capsicum* as ornamental plant is famous to be called as *cabai pelangi* or *rainbow cabai* in Indonesia.

4. Conclusion

The major *Capsicum* species commercially important in Indonesia are *C. annuum* and *C. frutescens*. Cultivars from *C. chinense* and *C. pubescens* were rarely found in the market and can be considered exotic and mostly cultivated for ornaments. *Capsicum* is profoundly intertwined and inseparable from Indonesian culture, from food and pharmaceuticals to ornaments with many variations of the products. The market is still exceptionally receptive to innovations to transform and include *Capsicums* in the products. The variety of *Capsicums* in Indonesia and the potential applications combined with prospective market also make the field of flavor and bioactive compounds of *Capsicums* in Indonesia attractive to be investigated.

Acknowledgements

We would like to express our gratitude to Ms. Zee Agustine, Mr. Irman F. Amrullah (PT Bimandiri Agrosedaya), Mr. Erick Tjiptadji, Ms. Ajeng Kinanti

Sugianto, Ms. Asih, Ms. Sarah Fathia, Ms. Cita (Cemilan Kabayan), Banua Cokelat, Ms. Yanti (RnR Coklat), and Prof. Muhammad Syukur (IPB University) for contributing their pictures in this chapter.

Author details

Cristofora Hanny Wijaya[1*], Manik Harda[2] and Bunga Rana[1]

1 IPB University, Bogor, Indonesia

2 Gadjah Mada University, Yogyakarta, Indonesia

*Address all correspondence to: channywijaya@apps.ipb.ac.id

References

[1] Yamamoto S, Djarwaningsih T, Wiriadinata H. History and distribution of *Capsicum chinense*. Tropical Agriculture and Development. 2014;**58**:94-101. DOI: 10.11248/jsta.58.94

[2] Ranutinoyo S. Cabai: aspek sejarah dan budaya [Internet]. 2015. Available from: https://www.kompasiana.com/tyo-setiadi/551b3685a33311b023b65cdc/cabai-aspek-sejarah-dan-budaya [Accessed: 12 May 2020]

[3] Lingga L. Health Secret of Pepper (Cabai). Jakarta: PT Elex Media Komputindo; 2012. pp. 1-13

[4] Djarwaningsih T. Review *Capsicum* spp: Asal, persebaran dan nilai ekonomi. Biodiversitas. 2005;**6**:292-296

[5] Thampi PSS. A glimpse of the world trade in *Capsicum*. In: De AK, editor. *Capsicum*: The Genus Capsicum. London: Taylor and Francis; 2003. pp. 16-24

[6] Wahyuni Y, Ballester A, Sudarmonowati F, Bino RJ, Bovy AG. Secondary metabolites of *Capsicum* species and their importance in human diet. Journal of Natural Products. 2013;**76**:783-793. DOI: 10.1021/np300898z

[7] Giuffrida D, Dugo P, Torre G, Bignardi C, Cavazza A, Corradini C, et al. Characterization of 12 *Capsicum* varieties by evaluation of their carotenoid profile and pungency determination. Food Chemistry. 2013;**140**:794-802. DOI: 10.1016/j.foodchem.2012.09.060

[8] Hasibuan HA, Meilano R. The utilization of red palm oil in preparation of stir-fry red Chilli sauce. Jurnal Teknologi Pertanian. 2018;**19**:95-106

[9] Adiyoga W. Faktor-faktor yang mempengaruhi perilaku dan keputusan konsumen untuk membeli kentang, bawang merah, dan cabai merah. Jurnal Hortikultura. 2011;**21**:280-294. DOI: 10.21082/jhort.v21n3.2011.p280-294

[10] Pinto CMF, dos Santos IC, de Araujo FF, da Silva TP. Pepper importance and growth (*Capsicum* spp.). In: do Rego FR, do Rego MM, Finger FL, editors. Production and Breeding of Chilli Peppers (*Capsicum* spp.). Switzerland: Springer; 2016. pp. 1-26. DOI: 10.1007/978-3-319-06532-8_1

[11] FAOSTAT United Nations Food and Agriculture Organization, Rome. FAO-STAT Online [Internet]. 2019. Available from: http://faostat.fao.org/default/aspx?lang-en [Accessed: 15 May 2020]

[12] Siregar EMB. Repository Universitas Sumatera Utara Lecture Paper on uji virulensi CMV asal Sumatera Utara pada tanaman cabai [Internet]. 2005. Available from: http://repository.usu.ac.id/handle/123456789/991 [Accessed: 13 May 2020]

[13] Aria P. Harga Cabai Naik, Menakar Dampak Banjir Terhadap Inflasi [Internet]. 2020. https://katadata.co.id/telaah/2020/01/20/harga-cabai-naik-menakar-dampak-banjir-terhadap-inflasi [Accessed: 12 May 2020]

[14] Zachariah TJ, Gobinath P. Paprika and Chili. In: Parthsarathy VA, Chempakam B, Zachariah TJ, editors. Chemistry of Spices. London: CABI; 2008. pp. 260-285. DOI: 10.1016/j.tifs.2009.10.002

[15] Ferniah RS, Pujiyanto S. Biodiversity of Indonesian red chilli (*Capsicum annuum* var. Longum) based on morphological characters. Advanced Science Letters. 2013;**23**:6462-6464. DOI: 10.1166/asl.2017.9654

[16] Simatupang S. Teknologi Budidaya Cabai Merah. Sumatera Utara: Balai

Pengkajian Teknologi Pertanian Sumatera Utara; 2012. pp. 1-19

[17] Berlian Z, Syarifah S, Sari DS. Pengaruh pemberian limbah kulit kopi (*Coffea robusta* L.) terhadap pertumbuhan cabai keriting (*Capsicum annum* L.). Jurnal Biota. 2015;**1**:22-32

[18] Harpenas A, Dermawan R. Budidaya Cabai Unggul. Penebar Swadaya: Jakarta; 2009

[19] Sumpena U. Penetapan kadar capsaicin beberapa jenis cabe (*Capsicum* sp.) di Indonesia. Media. 2013;**9**:9-16. DOI: 10.31942/md.v9i2.1328

[20] Kurniawati N. Sehat dan Cantik Alami Berkat Khasiat Bumbu Dapur. Mizan: Bandung; 2010

[21] Indonesian Agriculture Ministry. New varieties description. Jakarta: Germination Horticulture Directorate, Directorate General of Horticulture; 2011

[22] Manikharda TM, Arakaki M, Yonamine K, Hashimoto F, Takara K, Wada K. Influence of fruit ripening on color, organic acid contents, capsaicinoids, aroma compounds, and antioxidant capacity of Shimatogarashi (*Capsicum frutescens*). Journal of Oleo Science. 2018;**67**:113-123. DOI: 10.5650/jos.ess17156

[23] Nurjanah S, Sudaryanto Z, Widyasanti A, Pratiwi H. Antibacterial activity of *Capsicum annuum* L. oleoresin. Acta Horticulturae. 2016;**1125**:189-194. DOI: 10.17660/ActaHortic.2016.1125.23

[24] Murdiati G. Bumbu, Penyedap dan Penyerta Masakan Indonesia. Jakarta: PT Gramedia Pustaka Utama; 2013. pp. 7-10

[25] BPS (Badan Pusat Statistik). Produksi tanaman sayuran [Internet]. 2018. Available from: https://www.bps.go.id/subject/55/hortikultura.html#subjekViewTab3 [Accessed: 12 May 2020]

[26] Manikharda M, Takahashi M, Arakaki M, Yonamine K, Hashimoto F, Takara K, et al. Physical properties, flavor characteristics and antioxidant capacity of Shimatogarashi (*Capsicum frutescens*). Food Science and Technology Research. 2017;**23**:427-435. DOI: 10.3136/fstr.23.427

[27] Ashari S. Holtikultura Aspek Budaya. Jakarta: UI-Press; 2006

[28] Calzada TR, Qian M, Strid A, Neugart S, Schreiner M, Pacheco IT, et al. Effect of UV-B radiation on morphology, phenolic compound production, gene expression, and subsequent drought stress responses in chili pepper (*Capsicum annuum* L.). Plant Physiology and Biochemistry. 2018;**134**:94-102. DOI: 10.1016/j.plaphy.2018.06.025

[29] Nimrotham C, Songprakorp R, Thepa S, Monyakul V. Experimental research of drying red chili by two methods: Solar drying and low - temperature system drying. Energy Procedia. 2017;**138**:512-517. DOI: 10.1016/j.egypro.2017.10.237

[30] Zulkarni R, Aria M, Ningrum RM. Uji aktivitas antidiabetes ekstrak etanol buah cabai merah (*Capsicum annuum* L.) terhadap mencit putih jantan. SCIENTIA: Jurnal Farmasi dan Kesehatan. 2019;**9**:86-92. DOI: 10.36434/scientia.v9i1.221

[31] Pangastuti A, Wulandari AM, Amin AZ, Amin M. Mengungkap potensi senyawa alami dari cabai (*Capsicum annuum* L.) sebagai agen anti-autism melalui teknik reverse docking. Prosiding Semnas Hayati IV. Universitas Nusantara PGRI Kediri. 2016:123-129. DOI: 10.17605/OSF.IO/9TCG4

[32] Rahmi M, Aria M, Rahmi NA. Uji aktivitas antihiperkolesterolemia

ekstrak etanol buah cabai merah (*Capsicum annuum* L) terhadap mencit putih jantan. SCIENTIA : Jurnal Farmasi dan Kesehatan. 2017;7:105-112. DOI: 10.36434/scientia.v7i2.125

[33] Yang D, Luo Z, Ma S, Wong WT, Ma L, Zhong J, et al. Activation of TRPV1 by dietary capsaicin improves endothelium-dependent vasorelaxation and prevents hypertension. Cell Metabolism. 2010;**12**:130-141. DOI: 10.1016/j.cmet.2010.05.015

[34] Makmur D. Kitab Sakti Petani Cabai. Dadi Makmur: Jakarta; 2017

[35] Cahyono B. Cabai Paprika: Teknik Budidaya dan Analisis Usaha Tani. Jogjakarta: Penerbit Kanisius; 2003. pp. 13-28

[36] Wijayakusuma HMH. Ramuan Herbal Penurun Kolesterol. Pustaka Bunda, Depok: Depok; 2008. p. 72

[37] Syafriana V, Natasha N, Wahidin W. Uji aktivitas antibakteri ekstrak etanol buah paprika merah (*Capsicum annuum L.*) terhadap bakteri *Enterococcus faecalis*. Sainstech Farma. 2019;**1**:44-47

[38] Ogunruku OO, Oboh G, Passsamonti S, Tramer F, Boligon AA. *Capsicum annuum* var. grossum (bell pepper) inhibits β-secretase activity and β-amyloid$_{1\text{-}40}$ aggregation. Journal of Medicinal Food. 2017;**20**:124-130. DOI: 10.1089/jmf.2016.0077

[39] Manikharda. Physicochemical properties, flavor characteristics and biological functions of the chili pepper Shimatogarashi (*Capsicum frutescens*) [dissertation]. Japan: United Graduate School of Agricultural Sciences, Kagoshima University; 2018

[40] Sukrasno S, Kusmardiyani S, Tarini S, Sugiarso NC. Kandungan capsaicin dan dihidrocapsaicin pada berbagai buah *Capsicum*. Journal of Management Studies. 1997;**2**:28-34

[41] Kusnadi J, Andayani DW, Zubaidah E, Arumingtyas EL. Ekstraksi senyawa bioaktif cabai rawit (*Capsicum frutescens* L.) menggunakan metode ekstraksi gelombang ultrasonik. Jurnal Teknologi Pertanian. 2019;**20**:79-84. DOI: 10.21776/ub.jtp.2019.020.02.1

[42] Materska M, Perucka I. Antioxidant activity of the main phenolic compounds isolated from hot pepper fruit (*Capsicum annuum* L.). Journal of Agricultural and Food Chemistry. 2005;**53**:1750-1756. DOI: 10.1021/jf035331k

[43] Bley K, Boorman G, Mohammad B, McKenzie D, Babbar S. A comprehensive review of the carcinogenic and anticarcinogenic potential of capsaicin. Toxicologic Pathology. 2012;**40**:847-873. DOI: 10.1177/0192623312444471

[44] Kim CS, Kawada T, Kim BS, Han IS, Choe SY, Kurata T, et al. Capsaicin exhibits anti-inflammatory property by inhibiting IkB-a degradation in LPS-stimulated peritoneal macrophages. Cellular Signalling. 2003;**15**:299-306. DOI: 10.1016/S0898-6568(02)00086-4

[45] Backonja MM, Malan TP, Vanhove GF, Tobias JK. NGX-4010, a high-concentration capsaicin patch, for the treatment of postherpetic neuralgia: A randomized, double-blind, controlled study with an open-label extension. Pain Medicine. 2010;**11**:600-608. DOI: 10.1111/j.1526-4637.2009.00793.x

[46] Supendi D. Manfaat cabai rawit untuk kesehatan yang jarang diketahui [Internet]. 2020. Available from: https://www.harapanrakyat.com/2020/01/manfaat-cabai-rawit-untuk-kesehatan-ini-jarang-diketahui/ [Accessed: 12 May 2020]

[47] Yamamoto S, Djarwaningsih T, Wiriadinata H. *Capsicum pubescens* (Solanaceae) in Indonesia: Its history, taxonomy, and distribution. Economic Botany. 2013;**67**:161-170. DOI: 10.1007/s12231-013-9230-y

[48] Vebriansyah R. Tingkatkan Produktivitas Cabai. Jakarta: Penebar Swadaya Grup; 2018. p. 19

[49] Ranteallo AH. 2019. Cabe Katokkon Super Pedas. Available from: http://cybex.pertanian.go.id/mobile/artikel/85437/CABE-KATOKKON-SUPER-PEDAS/ [Accessed: 12 May 2020]

[50] Amaliah N. Penentuan kadar capsaicin menggunakan metode Kromatografi Lapis Tipis (KLT) pada cabe Kotokkon. JST (Jurnal Sains Terapan). 2018;**4**(1):49-56. DOI: 10.32487/jst.y4il.452

[51] Basoni S. Dari Dieng hingga Toraja, Ini 5 Jenis Cabai Super Pedas Indonesia [Internet]. 2019. Available from: https://food.detik.com/info-kuliner/d-4691602/dari-dieng-hingga-toraja-ini-5-jenis-cabai-super-pedas-indonesia/4/#news [Accessed: 12 May 2020]

[52] Pramudyani L, Sabran M, Noor A. Agronomic performance and nutrition content of Hiyung as local variety of cayenne pepper (*Capsicum frutescens*) at dry land and swamp land of South Kalimantan province. Buletin Plasma Nutfah. 2019;**25**:43-52. DOI: 10.21082/blpn.v25n1.2019.p43-52

[53] Kirnadi AJ, Zuraida A. Tingkat kelayakan usaha cabai hiyung (*Capsicum frutescens*) di beberapa luas lahan di desa Hiyung kecamatan Tapin Tengan kabupaten Tapin Kalimantan Selatan. Al-Ulum: Jurnal Sains dan Teknologi. 2019;**4**:33-40. DOI: 10.31602/ajst.v4i2.1977

[54] Dinas Pertanian Tanaman Pangan dan Hortikultura Kabuaten Tapin. Laporan Kinerja Instansi Pemerintah tahun anggaran 2015 [Internet]. 2016. Available from: https://docplayer.info/59016084 -Laporan -kinerja-instansi- pemerintah-lkip-tahun-anggaran- 2015- dinas-pertanian-tanaman-pangan- dan-hortikultura-kabupaten-tapin.html [Accessed: 19 May 2020]

[55] Hayati A, Hardarani N. Karakteristik lahan dan budidaya cabai rawit hiyung: Informasi dasar untuk peningkatan produksi cabai rawit hiyung di lahan rawa lebak. Prosiding – Seminar Nasional Lahan Basah. 2009;**4**(1):57-59

[56] Widiyastuti DA. Pengetahuan dan Sikap Petani Terhadap Hama Cabai Rawit Hiyung. Agrisains. 2015;**1**:63-70

[57] Sofiari E, Kirana R. Analisis pola segregasi dan distribusi beberapa karakter cabai. The Horticulture Journal. 2009;**19**:255-263

[58] Rahman F. Jejak Rasa Nusantara: Sejarah Makanan Indonesia. Gramedia Pustaka Utama: Jakarta; 2016. p. 23

[59] Widyanto *A. Indonesia* Punya 322 Jenis Sambal [Internet]. 2018. Available from: https://joss.co.id/2018/05/indonesia-punya-322-jenis-sambal/ [Accessed: 17 May 2020]

[60] Standar Nasional Indonesia SNI. Saus Cabe. In: SNI 01-2976-2006. Badan Standarisasi Nasional; 2006

[61] Wiranti EW, Hatmi RU, Raharjo HB, Widyayanti S. Teknologi Paska Panen dan Pengolahan Cabai. Yogyakarta: Balai Pengkajian Teknologi Pertanian Yogyakarta; 2011. pp. 9-20

[62] Kusuma ASW, Rosalina G. Analisis kadar kapsaisin dari ekstrak 'Bon Cabe' dengan menggunakan kromatografi cair kinerja tinggi (KCKT). Farmaka. 2016;**14**:11-18. DOI: 10.24198/jf.v14i2.9286

[63] Nazmudin A. Coklat Manis Sudah Biasa, ini Ada Coklat Manis Rasa Cabe [Internet]. 2016. Available from: https://wartakota.tribunnews.com/2016/10/16/cokelat-manis-sudah-biasa-ini-ada-cokelat-pedas-rasa-cabe [Accessed: 15 May 2020]

[64] Permatasari D. Kopi Cabe Rawit Pereda Flu [Internet]. 2017. https://www.cnnindonesia.com/tv/20170510101123-402-213713/kopi-cabe-rawit-pereda-flu [Accessed: 15 May 2020]

[65] Chrubasik S, Weiser T, Beime B. Effectiveness and safety of topical capsaicin cream in the treatment of chronic soft tissue pain. Phytotherapy Research. 2010;**24**:1877-1885. DOI: 10.1002/ptr.3335

[66] Altman RD, Aven A, Holmburg CE, Pfeifer LM, Sack M, Young GT. Capsaicin cream 0.025% as monotherapy for osteoarthritis: A double-blind study. Seminars in Arthritis and Rheumatism. 1994;**23**(6):25-33. DOI: 10.1016/S0049-0172(10)80023-X

[67] Ellison N, Loprinzi CL, Kugler J, Hatfield AK, Miser A, Sloan JA, et al. Phase III placebo-controlled trial of capsaicin cream in the management of surgical neuropathic pain in cancer patients. Journal of Clinical Oncology. 1997;**15**:2974-2980. DOI: 10.1200/JCO.1997.15.8.2974

[68] Tempo Administrator. Panasnya Koyo Cabe [Internet]. 1985. Available from: https://majalah.tempo.co/read/ekonomi-dan-bisnis/40088/panasnya-koyo-cabe [Accessed: 13 May 2020]

[69] Indosar. Koyo Cabe [Internet]. 2020. Available from: http://www.indosar.co.id/product-details/koyo-cabe.php [Accessed: 14 May 2020]

[70] Syukur M, Maharijaya A, Aisyah SI, Sukma D, Sulassih RAW, Istiqlal MR, et al. Varietas Cabai Hias Ayesha IPB. Horticulturae Journal. 2018;**2**:49-56. DOI: 10.29244/chj.2.1.49-56

[71] Djarwaningsih T. Pemanfaatan jenis-jenis cabai (*Capsicum* spp.) sebagai tanaman hias. Buletin Kebun Raya. 1983;**6**:45-52

[72] Syukur M, Maharijaya A, Aisyah SI, Sukma D, Sulassih RAW, Istiqlal MR, et al. Potensi keunggulan tanaman cabai Lembayung IPB sebagai varietas baru pada tanaman hias. Horticulturae Journal. 2018;**2**:54-61. DOI: 10.29244/chj.2.2.54-61

[73] Sulassih M, Syukur A, Sobir A, Maharijaya A, Hakim AR. Karakterisasi Lima Galur Cabai Hias dalam Rangka Pendaftaran Varietas Hasil Pemuliaan. Horticulturae Journal. 2017;**1**(1):26-33. DOI: 10.29244/chj.1.1.26-33

12

Vitamin E: Natural Antioxidant in the Mediterranean Diet

Samia Ben Mansour-Gueddes and Dhouha Saidana-Naija

Abstract

Oxidation has been related to several diseases in humans. Indeed, to protect the body from high free radical damages, organism requires natural resources of antioxidant compounds, such as phenols, tocopherols (α, β, γ, and σ) which have important roles in the cell antioxidant defense system. In Mediterranean areas, olive oils and pepper fruits are considered among the best foods in a diet, which keeps on attracting the interest of scientists due to the health benefits linked with its con-sumption. The Olive oil and pepper fruits are among the most consumed nutrients in the Mediterranean diet; their richness in naturally powerful antioxidants, such as alpha-tocopherols, polyphenols, carotenoïds, and capsaicinoïds (specific of capsicum species), and monounsaturated fatty acids in olive and seed pepper oils, constitutes good health protection against oxidative damages and inflammation. Also, these phytochemicals shield and prevent the human body from many diseases such as cardiovascular, coronary, Alzheimer's diseases, and cancers.

Keywords: tocopherols, antioxidants, *Olea europaea*, *Capsicum* sp., fruits, oils, Mediterranean areas

1. Introduction

In recent years, oxidation constitutes a major problem in human health. Oxygen is considered a vital element, but its instability has deleterious effects on the human body. At high concentration, free radicals cannot gradually be destroyed, their accumulation in the organism generate oxidative stress. This process can damage all cell structures as lipids, proteins, and DNA and trigger many human diseases, such as cancer, arteriosclerosis, and rheumatoid arthritis. Moreover, it may play a role in neurodegenerative diseases and aging processes [1]. Hence, to protect human organisms against reactive oxygen species (ROS) a request for external nutritional intake rich in antioxidants can assist in coping with this oxidative stress. Many studies accorded that dietary vitamin antioxidants and polyphenols have been explored extensively as an exogenous mechanism of defense against oxidative stress [1, 2].

The antioxidants exert their activity by scavenging the 'free-oxygen radicals' thereby giving rise to a fairly 'stable radical'. The human body produces an insufficient level of antioxidants which are essential for preventing oxidative stress. To protect against oxidant radicals, organism requires natural resources of antioxidant compounds from nutrients of different origins. These bioactive molecules play an important role in helping endogenous antioxidants for the neutralization of free radicals. Nutrient antioxidant deficiency is one of the causes of numerous chronic and degenerative pathologies [3].

It has been demonstrated that many vegetables, fruits, medicinal plants, and other foods contain compounds with bioactivity against oxidative stress. This activity has been attributed to vitamin C, vitamin E, α-tocopherol, β-carotene, and polyphenolic compounds [4, 5]. Therefore, research regarding natural antioxidants from foods and plants, particularly from folk medicinal plants, is receiving increas-ing attention around the world.

In Mediterranean areas, nutrition is specific to each country. The Mediterranean diet takes into account the various religious and cultural traditions, as well as the various national identities, the current needs of Mediterranean populations, respecting regional dietary variations. The Mediterranean model, qualified as a healthy lifestyle [6]. It is con-sidered as a model of sustainable nutrition [7] due to its richness in vegetables, fruits, in a quantity moderate fish, dairy, and meat products, condiments, and spices [8]. In fact, in Mediterranean countries, especially in North Africa, as Tunisia, the diet is based on olive oils, olive derivatives, cereals, Solanaceae species (pepper, tomatoes, potatoes), green vegetables, legumes, fresh and dried fruits. These sources of aliment are rich in macro and micro-nutrient such as fibers, antioxidants (vitamins, polyphenols, carotenoids), oligo-elements. In general, the Mediterranean diet is low in animal fat and fast sugar, but it is rich in fiber, omega 3, and antioxidants. The abundance of fresh fruits and vegetables and the use of olive oil instead of hard fats are key factors for which the Mediterranean diets are renowned. So many researches showed that Mediterranean eating habits appeared to meet all the criteria for a healthy and prudent diet [6, 9, 10].

In addition to macronutrients, humans need vitamins and minerals which are micronutrients required by the body to carry out a range of normal functions. However, these micronutrients are not produced in our bodies and must be derived from the food we eat. Vitamins, such as Vitamin A, B, C, E... are crucial for normal development. These micronutrients protect the organism against many diseases by their antioxidant property. In this context, this work aims to evaluate the richness in antioxidants, especially vitamin E, of the main Mediterranean nutriment olive oils, and pepper and their role in preventing many diseases.

2. Properties of vitamin E

2.1 Chemical properties

Vitamin E is composed of eight naturally isoforms, four tocopherols with the same molecular formula C28H48O2 (α-, β-, γ-, and σ-tocopherols), and four tocotrienols with a molecular formula C26H38O2 (α-, β-, γ-, and σ-tocotrienols) (**Figure 1**). These molecules are synthesized by photosynthetic organisms including plants, algae, and cyanobacteria, from homogentisic acid and phytyl-diphosphate or farnesyl-diphosphate reaction in plastid membranes [11]. All homologs are derivatives of 6-chromanol and differ in the number and position of methyl groups on the ring structure. The four tocopherol homologs have a saturated 16-carbon phytyl side chain. Whereas, the tocotrienols homologs have three double bonds on the side chain. The tocopherols and tocotrienols have the same basic chemical structure; the main difference is in the saturation of the aliphatic side chain attached to the chromanol ring [12, 13] (**Figure 1**). The various isoforms are not interchangeable and only α-tocopherol meets the human vitamin E requirements [14].

The reaction of aromatic chromanol ring with phytyl-diphosphate or farnesyl-diphosphate, in plastid membranes, synthesize different isomers of tocopherol or tocotrienol.

Vitamin E is a fat-soluble vitamin that exists in different chemical forms. The most active compound is alpha-tocopherol which existed in natural and synthetic forms.

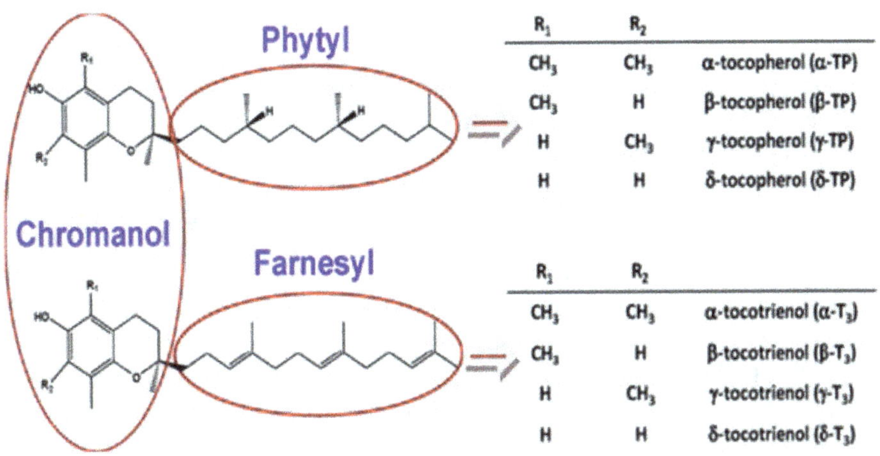

Figure 1.
Chemical structure of different isoforms of vitamin E [13].

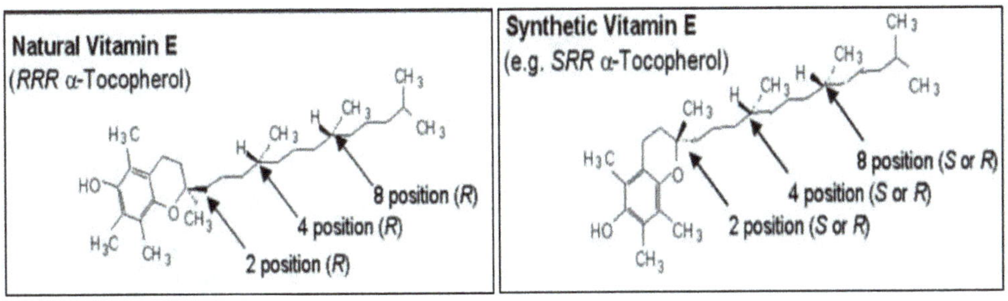

Figure 2.
Natural and synthetic Sterio-isoforms of vitamin E [14].

The natural vitamin E (RRR-α-tocopherol or D-α-tocopherol) consists of a single stereoisomer. While, synthetic vitamin E is a mixture of eight stereoisomers (RRR, RSR, RSS, RRS, SRR, SRS, SSR, SSS) distributed equally (**Figure 2**). Only one of them (1/8th) has a molecular structure identical to that of the natural vitamin. According to many studies, the natural vitamin E is twice as powerful and fixed twice as good as the synthetic version. This means that natural vitamin E reaches the blood and organs at least twice as good as synthetic ones [13, 14].

The stereoisomer S and R are the spatial arrangements of the alpha-tocopherol; The RRR refers to R at the 2,4 and 8 positions, hence RRR-α-tocopherol which correspond to Natural vitamin E. The synthetic Vitamin E agree to eight stereoisomers S and/or R at positions 2, 4 and 8.

2.2 Biological properties of vitamin E in the human body

Vitamin E is considered a natural phytochemical that is frequently associated with human health [15]. This vitamin is an example of a phenolic antioxidant; it serves as an antioxidant and protects membrane lipids from oxidative degeneration [16]. Therefore, both lipophilicity and membrane localization of vitamin E explain its antioxidant property. In this context, the incorporation of vitamin E into the cell membrane explained their major biologic role to protect polyunsaturated fatty acids (PUFAs) and other components of cell membranes and low-density lipoprotein (LDL) from oxidation by free radicals. Via their localization, within the phospholipid bilayer of cell membranes; It is particularly effective in preventing lipid peroxidation, a series of chemical reactions involving the oxidative deterioration of PUFAs [14].

Vitamin E under the term α-tocopherol is a powerful biological antioxidant. It is the major lipid-soluble component in the cell antioxidant defense system and is exclusively obtained from the diet. Among the eight isomers, The RRR-α-tocopherol is the most isoform of vitamin E that is essential for humans and is preferentially retained within the organism [14, 17, 18]. This is explained in part by the specific selection of RRR-α-tocopherol by the α-tocopherol transfer protein and in part by its low rate of degradation and elimination compared with the other vitamers, especially tocotrienols, which are rapidly metabolized and excreted similarly as other xenobiotics [14]. Also, [19] mentioned that γ-Tocopherol is slightly less efficient than α-tocopherol as a scavenger of oxygen radicals, but it is an efficient scavenger of reactive nitrogen species due to the unsubstituted 5-position on the chromanol ring. This isomer of Tocopherol is present in significant amounts in the human diet especially in several widely consumed vegetable oils [14].

This form is considered the most important fat-soluble antioxidant in humans metabolizing peroxyl radicals [19]. Such molecules readily donate the hydrogen from the hydroxyl (-OH) group on the ring structure to free radicals, which then become unreactive. On donating the hydrogen, the phenolic compound itself becomes a relatively unreactive free radical because the unpaired electron on the oxygen atom is usually delocalized into the aromatic ring structure thereby increasing its stability [14].

The potent lipid-soluble antioxidant property of α-tocopherol is to maintain the integrity of long-chain polyunsaturated fatty acids in the membranes of cells and thus maintain their bioactivity [20]. The α-tocopherol protects the peroxidation of unsaturated fatty acids of the cell membrane. When peroxyl radicals ($ROO^{\bullet}$) are formed, these react 1000-times faster with vitamin E (Vit E-OH) than with polyunsaturated fatty acids (PUFA: ROOH) [21]. The hydroxyl (OH) group in the chromanol head of α-tocopherol can donate hydrogen to scavenge lipid peroxyl radicals ($ROO^{\bullet}$) generated from the peroxidation of the lipids to form the corresponding lipid hydroperoxide and the tocopheryl radical (Vit E-$O^{\bullet}$). The tocopheryl radical (Vit E-$O^{\bullet}$) reacts with vitamin C, thereby oxidizing the latter and returning

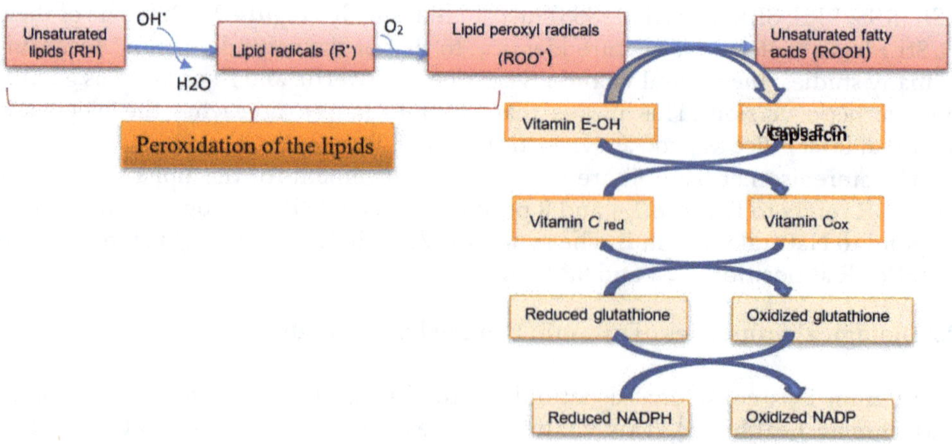

Figure3.
The antioxidant property of Vitamin E and its regeneration by other antioxidants "Vitamin E recycling". Vitamin E-OH: alpha-tocopherol; Vitamin E-O.: tocopheryl radical; NADP: nicotinamide adenine diphosphate; NADPH: reduced NADP. The peroxidation of unsaturated lipids leads to forming lipid peroxyl radicals (ROO·). α-tocopherol easily diffuses into cell membranes, due to its lipophilic nature, and scavenges rapidly, with a hydroxyl group, the lipid peroxyl radicals and protects polyunsaturated fatty acids from lipid peroxidation. The redox reaction between tocopherol and harmful lipid peroxide radicals leads to forming neutral lipid hydroperoxide and an unreactive vitamin E radical (Vitamin E-O.). The presence of other antioxidants, such as vitamin C, is required to regenerate the antioxidant capacity of α-tocopherol.

vitamin E to its reduced state. The presence of other antioxidants such as vitamin C is required to regenerate the antioxidant capacity of α-tocopherol (**Figure 3**).

The interaction of vitamins E and C has led to the idea of "vitamin E recycling", where the antioxidant function of oxidized vitamin E is continuously restored by other antioxidants (**Figure 3**).

3. Role of vitamin E against diseases

The Mediterranean diet (MedDi) is characterized by a high content of bioactive phytochemicals especially the antioxidants such as polyphenols, carotenoids, vitamins C and E which are key components of many plant foods. These bioac-tive compounds in (MedDi) have a particular interest in the prevention of many diseases such as cancer, cardiovascular diseases, etc. [22]. Also, the major dietary sources of vitamin E are fruits, vegetables, nuts, and oils. Vitamin E is known to inhibit lipid peroxidation eventually protecting DNA from damage involved in the pathogenesis of cancer [23].

Antioxidants from our diet play an important role in helping endogenous antioxidants for neutralizing free radical species. Thus, free radicals are involved in some diseases including tumor inflammation, hemorrhagic shock, atherosclerosis, diabetes, infertility, gastrointestinal ulcerogenesis, asthma, rheumatoid arthritis, cardiovascular disorders, neurodegenerative diseases, etc. [24, 25].

By its natural antioxidant property, vitamin E plays a key role in maintaining health and preventing many chronic and degenerative diseases [26]. In fact, vitamin E could help avoid or delay coronary heart disease, it could also prevent atherosclerosis by inhibits or reduces the oxidation of low-density lipoprotein (LDL) cholesterol which is associated with the development of atherosclerosis [27]. Also, the supplement of this nutrient plays a cardioprotective role and decreases cardiovascular events [28, 29], and avoids the formation of blood clots that could lead to a heart attack or venous thromboembolism [30]. Nutrient antioxidant deficiency is one of the causes of numerous chronic and degenerative pathologies [1].

Among the different forms of vitamin E, alpha-tocopherol constitutes the most biologically active form and is preferentially absorbed and retained in the body. It has anti-inflammatory, antiplatelet, and vasodilator properties with which vitamin E enhances the immune system presents the capacity to promote health, prevents and treats many diseases [29, 31]. Also, due to its natural properties to be fat-soluble and to incorporate into biological membranes, alpha-tocopherol prevents protein oxidation and inhibits lipid peroxidation, thereby maintaining cell membrane integrity and protecting the cell against damage [32]. Alpha-tocopherol also modulates the expression of various genes, plays a key role in neurological function, inhibits platelet aggregation, and enhances vasodilatation. Many researches showed that the supplementation of vitamin E (200–400 mg/day) may be suitable to moderate some aspects of degenerative diseases such as Parkinson's disease, reduce tissue injury arising from ischaemia and reperfusion during surgery, delay cataract development, and improve mobility in arthritis sufferers [33].

4. Vitamin E in olive trees

4.1 Importance of vitamin E in olive oil

Olive fruits and olive oils are considered excellent Mediterranean nutriments. Their consumption in the Mediterranean diet (MedDi) constitutes the cause of many health-promoting effects. The olive oils are characterized by their richness in oleic acid, vitamin E, polyphenols, and some other minor components some of which are known to be anti-inflammatory, make it the model functional food [34]. Olive oils, virgin, and extra virgin are a symbol of the Mediterranean Diet. Alpha-tocopherol was the most abundant tool and detected in all the studied olive oil samples [35]. Many research proved that the levels of tocopherols in olive oils are high variety and geographic areas-dependent.

The alpha-tocopherol is more active than others β > γ > δ against free radicals. It protects free fatty acids from peroxidation. The tocopherol radicals are resonance stabilized within the chromanol ring and do not propagate the chain reactions or are rapidly recycled back to the corresponding tocopherol, allowing each tocopherol to participate in many peroxidation chain-breaking events. One tocopherol molecule can protect about 103–108 polyunsaturated fatty acids at low peroxide values [36]. According to [37], α-tocopherol represented almost 95% of total tocopherols and their contribution is greater than the rest of tocopherols; their content in virgin olive oils varies from 97 to 785 mg/kg. In fact, α-tocopherol concentration ranges from 170 to 485 mg/kg in Spanish varieties [37]; 160 to 428 mg/kg in the

Countries	Varieties	Total Tocopherol (mg/kg)	α- tocopherol (mg/kg)	References
Spain	Arbequina	371 ± 16	373 ± 16	[37]
	Morisca	501 ± 26	485 ± 27	
	Picual	355 ± 19	346 ± 20	
	Manzanilla-cacerena	336 ± 31	333 ± 31	
	Corniche	366 ± 31	366 ± 31	
	Verdial de Badajoz	292 ± 23	311 ± 22	
	Carrasquena	310 ± 58	280 ± 56	
Greece	Koroneiki	121 ± 22	117 ± 21	[42]
	Kolovi	123 ± 28	110 ± 15	
Turkey	Ayvalik	183.27 ± 15.5	180.43 ± 15.17	[35]
	Domat	160.78 ± 18.7	106.8 ± 18.27	
	Gemlik	114.87 ± 9.73	112.59 ± 9.49	
Italy	Leccino	455.25 ± 4.4	405.6 ± 4.6	[43]
	Frantoio	270.7 ± 1.7	230.0 ± 1.6	
Tunisia	Chetoui	405.65 ± 4.17	385.35 ± 2.48	[44]
	Chemlali	199	184	[44]
	Chemlali Sfax	467	425	[45]
	Chemlali Zarzis	400	374	[45]
	Oueslati	204	185	[45]
	Sayali	282	264	[45]
	Zalmati	351	336	[45]
	Zarrazi	208	193	[46]
	Meski	-	74.6 ± 4.6	[47]
	Neb Jmal	-	232.29 ± 2.00	[47]
	El Hor	-	335.27 ± 1.16	[47]
	Jdallou	-	364.23 ± 3.30	[47]
Maroc	Picholine marocaine	311 ± 11.4	272.0 ± 8.0	[48]
Algeria	Chemlal Bordj	202.35	193.55	[49]
	Arima	188.55	179.72	
	Chemlal Zenata	240.1	228.11	
	Chemlal (SBA)	215.6	202.9	
	Sigoise Sebra			

Table 1.
Total tocopherols and α-tocopherol composition in olive oils of different varieties from Mediterranean countries.

Argentinean oils [38]; 98–370 mg/kg in the Greek oils [37], 97–403 mg/kg in oils from Turkey [39], 120–478 mg/kg in oils from Tunisia [40], and 138–298 mg/kg in the Portuguese oils [41] (**Table 1**).

4.2 Effect of olive oils in many diseases

Olive oil is the main source of fats in Mediterranean diets. This type of diet has often been associated with improving the resistance to certain diseases, includ-ing cardiovascular disease and illness degenerative. Many scientific studies have focused on the nutritional aspect of olive oil to understand the mechanisms of this phenomenon. The first explanation is its specific fatty acid composition. The proportion of saturated fatty acids is very low (14%); while the majority of mono-unsaturated fatty acids (MUFA) is oleic acid. Essential polyunsaturated fatty acids (PUFA) are also present in interesting proportions in the oil. MUFA supplement allows to increase the resistance of LDL to oxidation [50], thus reducing one of the factors that can cause coronary heart disease [9].

The Extra Virgin Olive Oil (EVOO) is one of the most important health-protective foods in the Mediterranean diet [51]. The high-quality EVOO is considered as a true pharm-food. This oil contains a relevant concentration of efficient chemo-preventive molecules, including Tyrosol, hydroxytyrosol, tocopherols (vitamin E), β-carotene, and phenolic compounds [51, 52]. [53] showed the ability of VOO phenolic compounds to shield lipoproteins from oxidation and to reduce systolic blood pressure in hypertensive individuals. These antioxidants compounds are thought to be beneficial to protect against neurodegenerative diseases and cardiovascular diseases [54]. Also, [27] suggested that the antioxidants compounds in EVOO can prevent and treat cancers, diabetes, neurodegenerative diseases, inflammation, and aging. They have an antimicrobial property and also play an important role in strengthening the immune system and protecting certain tissues and organs from damage. The presence of phenolic compounds and tocopherols in Extra Virgin Olive Oil (EVOO) protects the unsaturated fatty acids from peroxidation, thus contrib-utes to the stability of cellular brain structures [51, 55] and it has beneficial effects on learning and memory [55]. The phenolic compound and tocopherols have often been linked to reducing the risk of cognitive decline and are essential for proper brain function.

5. Impact of pepper composition in human health

5.1 Importance of pepper in Mediterranean diet

Pepper is a very important vegetable worldwide and has economic and agro-food importance in many countries. In the Mediterranean, pepper is cultivated in the warm regions particularly in Tunisia where its cultivation has spread due to its strong uses in Tunisian cuisine. Pepper fruits were appreciated and consumed mostly as fresh food or dried as a spice. The nutritional contribution due to the presence of beneficial healthy-related compounds, Pepper is among the most fascinating and consumed spice foods, largely appreciated for its flavor, high nutritional and health contribution to human diets [56].

Pepper is a usual part of a traditional Mediterranean diet. Hot peppers are intensively used as food additives for their pungency, aroma, and color [57]. Their consumption is nutritionally valuable and also contains ingredients that promote health. The presence of phytochemicals and antioxidants in fruits increases its importance in controlling diseases to protect the human body from the harmful effects of free radicals [58]. Therefore, integrating a pepper-rich diet in our daily meals can prevent cardiovascular diseases, could help in fighting blood cholesterol levels, and can have, by capsaicin, an antidiabetic activity [58, 59].

5.2 Antioxidants in pepper fruits

Pepper fruits are recognized for their richness in phytochemicals and antioxidants with high nutritional value. The fruits are considered an excellent source of macro and micro-nutrients such as provitamin A, vitamins C and E, carotenoïds, capsaicinoïds, minerals, polyphenols, phytosterol, metabolites with famous antioxidant properties that positively affect human health [60–62]. These phytochemicals are influenced by a variety of peppers and environmental factors. The analysis of 23 accessions of peppers, collected from multiple Peruvian locations, showed that the tocopherols varied strongly from 0.23 to 29.1 mg/100 g, the total polyphenols between 0.97 and 2.77 g gallic acid equivalents (GAE)/100 g and the concentra-tions of capsaicinoïds range from 1.0 mg/100 g to 1515.5 mg/100 g (GAE)/100 g [63]. So, the consumption of pepper fruits and integrating a pepper-rich diet into our daily meals can be helpful in the continuing quest to combating micronutrient deficiency [64].

Peppers are considered one of the best sources of natural vitamin E and C. Many studies showed that the level of α-tocopherol in dry red pepper powder is similar to those in spinach and asparagus and four-fold higher than that in dry tomatoes [64]. The recommended daily intake of vitamin E was 15 mg/day of α-tocopherol for both women and men. Pepper fruits can supply above 100% α-tocopherol per 100 g serving depending on the cultivar [61]. Also, the red pepper seed oils showed a high antioxidant capacity due to their richness in bioactive phytochemical compounds such as polyphenols, carotenoids, tocopherols,

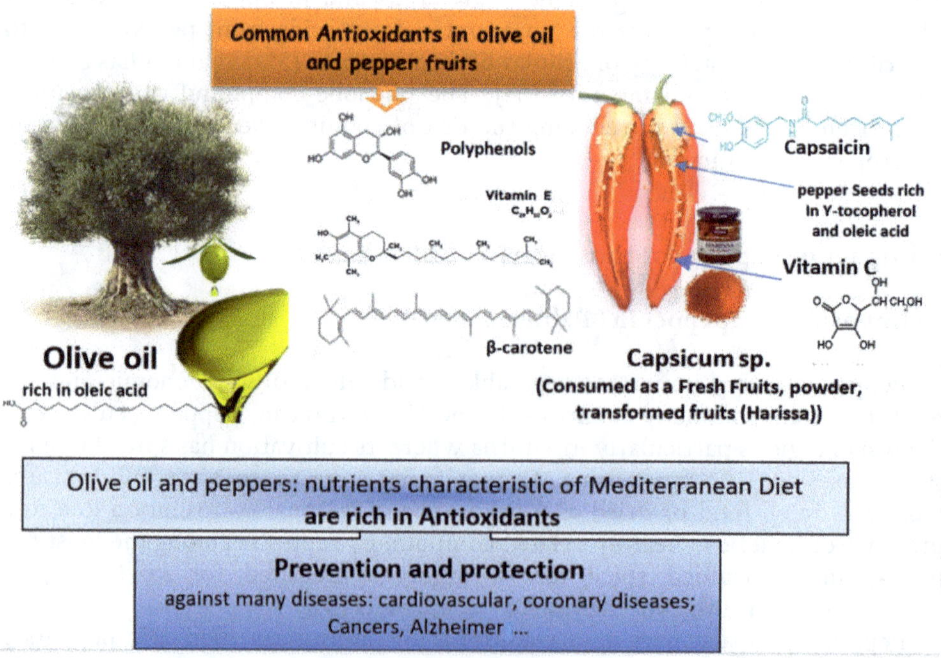

Figure4.
Graphical abstract of the main antioxidants commonly present in olive oils and pepper fruits, characteristic of the Mediterranean diet, such as phenolic compounds, carotenoids, tocopherols and oleic acid. The capsaicin is specific to capsicum species, which have anti-inflammatory properties.

phytosterols, and unsaturated fatty acids especially the linoleic acids which are higher than those of oleaginous seed oils [62]. It has been reported that polyphenols and tocopherol were the predominant antioxidant compounds in red pepper seed oils; Which γ-tocopherol was the main tocopherol at 278.65 mg/100 g seed oil, followed by alpha-tocopherol and delta-tocopherol [62]. The ratio between α- and γ-tocopherol depends on the number of seeds in the chili powder. As a matter of fact, the amount of α-tocopherol in the pericarp is higher, however, γ-tocopherol is more dominating in the seeds [63].

Hot pepper fruits are rich in capsaicinoïds, unique compounds of Capsicum species, which are responsible for the pungency. Capsaicin is widely influenced by the variety and maturity stages and by environmental factors [65]. These alcaloïds have antioxidant, anti-inflammatory, analgesic properties and are characterized by great medical and pharmacological values [58, 66]. These molecules also have a therapeu-tic effect as a neuropharmacological tool. Their effect in the treatment of painful conditions has been evaluated, such as rheumatic diseases [66]. Many recent studies have shown the effective treatment of capsaicinoïds for several sensory nerve fiber disorders, including arthritis and human immunodeficiency virus [67]. In this con-text, the proposed diet rich in pepper fruits can be considered an excellent strategy to improve the nutritional value of the population due to its high antioxidants and phenolic compound content.

The pepper fruits and olive oil constitute an excellent nutrient, for their richness in antioxidants (**Figure 4**). Since these Mediterranean foods have an important effect on human health, it is encouraged, around the world, to consume them.

6. Conclusion

The Mediterranean diet is rich in nutrients that have antioxidant properties. Particularly olive oil and pepper fruits constitute the most abundant and consumed vegetable nutriments in MedDi areas. Their richness in polyphenols, tocopherols, carotenoids, chlorophylls, unsaturated fatty acids, olive oils constitute a health treasure. These minor components are known to prevent and protect the human organism against many diseases such as cardiovascular, coronary diseases; also, some of these antioxidants have anti-inflammatory action, make it the model functional food. Pepper fruits are mostly consumed by the Mediterranean population as traditional spices and food products. Fruits are characterized by a means of antioxidants such as vitamin A alpha and gamma tocopherols, vitamin C, capsaicinoïds, polyphenols. The consumption of both nutriments rich in natural powerful antioxidants, such as tocopherols and polyphenols, constitutes a good strategy for reducing oxidative damage and to improve the health state of the human body, and preventing it from diseases.

Acknowledgements

This work was supported by the Laboratory of Amelioration of the Olive Tree Productivity and Product Quality. The author is grateful to all the members of the olive Tree institute.

Conflict of interest

The authors declare no conflict of interest for this chapter.

Author details

Samia Ben Mansour-Gueddes* and Dhouha Saidana-Naija
Olive Tree Institute, Sousse, Tunisia

*Address all correspondence to: bm_samiatn@yahoo.fr

References

[1] Pham-Huy LA, He H and Pham-Huy C. Free Radicals, Antioxidants in Disease and Health. Int J Biomed Sci. 2008; 4(2): 89-96.

[2] Billingsley H and Carbone S. The antioxidant potential of the Mediterranean diet in patients at high cardiovascular risk: an in-depth review of the Predimed. Nutrition and Diabetes. 2018; 8:13. DOI 10.1038/s41387-018-0025-1

[3] Lobo VA, Patil A, Phata K, and Chandra N. Free radicals, antioxidants and functional foods: Impact on human health. Pharmacognosy Reviews. 2010; 4 (8): 118-126. DOI:10.4103/0973-7847.70902

[4] Krishnaiah D, Sarbatly R and Nithyanandam R. A review of the antioxidant potential of medicinal plant species. Food and Bioproducts Processing. 2011; 89: 217-233.

[5] Moure A, Jose M, Daniel C, Franco J, Domínguez, M, Sineiro J, Domínguez H, Núñez JM, Parajó JC. Natural antioxidants from residual sources. Food Chemistry. 2001; 72 (2):145-171. DOI: 10.1016/S0308-8146(00)00223-5

[6] El-Sabban F. The antioxidant advantage of the Mediterranean diet in cardiovascular disease. Nutrition and Dietary Supplements. 2014; 6:35-40. DOI:10.2147/NDS.S59848

[7] González-Turmo I. Chapter 5: The Mediterranean Diet: consumption, cuisine and food habits. In: MediTERRA-CIHEAM, SciencesPo Les Presses Paris. 2012, p. 115-132. DOI: 10.3917/scpo.chea.2012.02

[8] Sikalidis AK, Kelleher AH and Kristo AS. Mediterranean Diet. Encyclopedia. 2021; 1: 371-387. DOI:10.3390/encyclopedia1020031

[9] Gerber M and Hoffman R. The Mediterranean Diet: Health, sciences and society. British Journal of Nutrition. 2015; 113, S4-S10. DOI:10.1017/S0007114514003912

[10] Boskou D. Chapter 13: Mediterranean food products: research and development. In: MediTERRA-CIHEAM, Presses de Sciences Po editors, 2012 p. 265-282. DOI:10.3917/scpo.chea.2012.02

[11] Fritsche S, Wang X and Jung C. Recent Advances in our Understanding of Tocopherol Biosynthesis in Plants: An Overview of Key Genes, Functions, and Breeding of Vitamin E Improved Crops, Antioxidants. 2017; 6(99):1-18. DOI: 10.3390/antiox6040099

[12] Rizvi S, Raza ST, Ahmed F, Ahmad A, Abbas S, and Mahdi F. The Role of Vitamin E in Human Health and Some Diseases. Sultan Qaboos Univ Med J. 2014; 14(2): e157–e165.

[13] Springett GM, Husain K, Neuger A, Centeno B, Chen DT, Hutchinson TZ, Lush RM, Sebti S, Malafa MP. A Phase I Safety, Pharmacokinetic, and Pharmacodynamic Presurgical Trial of Vitamin E δ-tocotrienol in Patients with Pancreatic Ductal Neoplasia. E BioMedicine. 2015; 2(12):1987-1995. DOI: 10.1016/j.ebiom.2015.11.025.

[14] Brigelius-Flohe R. Bioactivity of vitamin E. Nutr. Res. Rev. 2006, 19: 174-186. DOI: 10.1017/S0954422407202938

[15] Epriliati I and Ginjom I. Chapter 9: Bioavailability of Phytochemicals. In: Venketeshwer R Editor, Phytochemicals - A Global Perspective of Their Role in Nutrition and Health. InTechOpen; 2012 401-428. DOI:10.5772/26702

[16] Dhalaria R, Verma R, Kumar D, Puri S, Tapwal A, Kumar V, Nipovimova

E and Kuca K. Bioactive Compounds of Edible Fruits with Their Anti-Aging Properties: A comprehensive Review to Prolong Human Life. Antioxidants. 2020; 9 (1123): 1-38. doi:10.3390/antiox9111123

[17] Clarke MW, Burnett JR and Croft KD. Vitamin E in Human Health and Disease. Critical Reviews in Clinical Laboratory Sciences. 2008; 45(5): 417-450.

[18] World Health Organization/Food and Nutrition Division (FAO/WHO). Vitamin E: human vitamin and mineral requirements, FAO Rome. 2001; Pp. 291-325.

[19] Raederstorff D, Wyss A, Calder PC, Weber P and Eggersdorfer M. Vitamin E function requirements in relation to PUFA. British Journal of Nutrition. 2015; 114: 1113-1122.

[20] Traber MG and Atkinson J. Vitamin E, antioxidant and nothing more. Free radical biology and medicine. 2007; 43 (1), 4-15.

[21] Buettner GR. The pecking order of free radicals and antioxidants: Lipid-peroxidation, α-tocopherol, and ascorbate. Archives of Biochemistry and Biophysics. 1993; 300: 535-543. DOI:10.1006/abbi.1993.1074

[22] Lopez-Guarnido O, Urquiza-Salvata N, Saiza M, Lozano-Paniaguaa D, Rodrigo L, Gelerc Pascual M, Lorentea J A, Alvarez-Cubero M J and Rivas A. Bioactive compounds of the Mediterranean diet and prostate cancer. The aging male. 2018; 21(4): 251-260.

[23] Wright ME, Weinstein SJ, Lawson KA, et al. Supplemental and dietary vitamin E intakes and risk of prostate cancer in a large prospective study. Cancer Epidemiol Biomarkers Prev. 2007;16: 1128-1135.

[24] Chen FW, Shieh P, Kuo D, Hsieh C. Evaluation of the antioxidant activity of *Ruellia tuberosa*. Food Chemistry. 2006; 94:14-18.

[25] Uddin SN, Akond MA, Mubassara S, Yesmin MN. Antioxidants and antibacterial activities of *Trema cannabina*. Middle East Journal of antimicrobial Agents. 2008; 14: 327-335.

[26] Doughari JH. Chapter 1: Phytochemicals: Extraction Methods, Basic Structures and Mode of Action as Potential Chemotherapeutic Agents. In: Venketeshwer R Edition, Phytochemicals - A Global Perspective of Their Role in Nutrition and Health. IntechOpen, 2012; 1-34. DOI: 10.5772/26052

[27] Benlemlih M and Ghanem J. Polyphenols d'huile d'olive, tresor santé ! Polyphenol aux actions antioxydantes, anti-inflamatoires, anticancereuses, anti-veillissement et protectrices cardio-vasculaires. Deuxième editions, Medicatrix, Belgique, 2016; 208. ISBN 978-2-87211-159-6.

[28] Sesso HD, Buring JE, Christen WG, et al. Vitamins E and C in the prevention of cardiovascular disease in men: the Physicians' Health Study II randomized controlled trial. JAMA. 2008; 300: 2123-2133.

[29] Ziegler M, Wallert M, Lorkowski S and Peter K. Cardiovascular and Metabolic Protection by Vitamin E: A Matter of Treatment Strategy? Antioxidants (Basel). 2020; 9(10): 935. DOI: 10.3390/antiox9100935

[30] Jaffe R. Cardioprotective nutrient. In: Watson RR and Preedy VR Editors, Bioactive food as dietary interventions for cardiovascular disease. 1st edition, Elsevier, 2013; p. 103-120.

[31] Blumberg JB. Vitamin E supplementation and heart disease. Nutrition in Clinical Care. 2002;

5(2):50-55. DOI:10.1046/j.1523-5408.2002.00003.x

[32] Leskovec J, Levart A, Nemec Svete A, Perić L, Đukić Stojčić M, Žikić D, Salobir J, Rezar V. Effects of supplementation with α-tocopherol, ascorbic acid, selenium, or their combination in lin seed oil-enriched diets on the oxidative status in broilers. Poult. Sci. 2018; 97: 1641-1650

[33] Packer L. Vitamin E: Biological activity and Health Benefits: Overview. In: Packer L, Fuchs J, Dekker M editors, Vitamin E in health and disease. 1993; p. 977-982.

[34] Perona JS and Botham KM. Olive Oil as a Functional Food: Nutritional and Health Benefits. In: Aparicio, Harwood J. Handbook of Olive Oil, Springer, 2nd editions. 2013; pp. 677-714|.

[35] Dagdelen A, Tumen G, Ozcan MM and Dundar E. Determination of tocopherol contents of some olive varieties harvested at different ripening periods. Natural Product Research. 2012; 26 (15), 1454-1457. DOI:10.1080/14786419.2011.605364.

[36] Špika MJ, Kraljić K and Škevin D. Tocopherols: Chemical Structure, Bioactivity, and Variability in Croatian Virgin Olive Oils. In: Dimitrios Boskou and Maria Lisa Clodoveo editors, IntechOpen, DOI: 10.5772/64658.

[37] Franco MN, Galeano-Díaz T, Sánchez J, De Miguel C and Martín-Vertedor D. Total Phenolic Compounds and Tocopherols Profiles of Seven Olive Oil Varieties Grown in the South-West of Spain. Journal of Oleo Sciences. 2014; 63, (2) 000-000. DOI: 10.5650/jos.ess13098

[38] Ceci LN and Carelli AA. Characterization of Monovarietal Argentinian Olive Oils from New Productive Zones. Journal of the American Oil Chemist's Society. 2007; 84(12): 1125-1136. DOI:10.1007/s11746-007-1140-7.

[39] Arslan D, Karabekir Y, Schreiner M. Variations of phenolic compounds, fatty acids and some qualitative characteristics of Sarıulak olive oil as induced by growing area. Food Res Int. 2013; 54 (2):1897-1906. DOI:10.1016/j.foodres.2013.06.016.

[40] Baccouri O, Guerfel M, Baccouri B, Cerretani L, Bendini A, Lercker G, Zarrouk M, Daoud B, Miled D. Chemical composition and oxidative stability of Tunisian Monovarietal virgin olive oils with regard to fruit ripening. Food Chemistry. 2008; 109(4):743-754. DOI:10.1016/j.foodchem.2008.01.034

[41] Matos LC, Cunha SC, Amaral JS, Pereira JA, Andrade PB, Seabra RM, Oliveira BP. Chemometric characterization of three varietal olive oils (Cvs. Cobrançosa, Madural and Verdeal Transmontana) extracted from olives with different maturation indices. Food Chem. 2007;102(1):406-414. DOI:10.1016/j.foodchem.2005.12.031

[42] Mikrou T, Pantelidou E, Parasyri N, Papaioannou A, Kapsokefalou M, Gardeli C and Mallouchos A. Varietal and Geographical Discrimination of Greek Monovarietal Extra Virgin Olive Oils Based on Squalene, Tocopherol, and Fatty Acid Composition. Molecules, 2020; 25 (3818): 1-14. DOI:10.3390/molecules25173818

[43] Najafi V, Barzegar M, Sahari M. Physicochemical Properties and Oxidative Stability of Some Virgin and Processed Olive Oils. Journal of Agricultural Science and Technology. 2015; 17(4): 847-858

[44] Ben Temime S, Baccouri B, Taamalli W, Abaza L, Daoud D, Zarrouk M. Location effects on oxidative stability of chétoui virgin olive

oil. Journal of Food Biochemistry. 2006; 30 (6): 659-670. DOI: 10.1111/j.1745-4514.2006.00086.x

[45] Laroussi-Mezghani S, Le Dréau Y, Molinet J, Hammami M, Grati-Kamoun N, Artaud J. Biodiversity of Tunisian virgin olive oils: varietal origin classification according to their minor compounds. Eur. Food Res. Technol. 2016; 242:1087-1099.

[46] Sakouhi F, Harrabi S, Absalon C, Sbei K, Boukhchina S, Kallel H. α-Tocopherol and fatty acids contents of some Tunisian table olives (*Olea europaea* L.): Changes in their composition during ripening and processing. Food Chemistry. 2008;108: 833-839.

[47] Krichene D, Taamalli W, Daoud D, Salvador MD, Fregapane G, Zarrouk M. Phenolic compounds, tocopherols and other minor components in virgin olive oils of some Tunisian varieties. Journal of Food Biochemistry. 2007; 31 (2): 179-194. DOI:10.1111/j.1745-4514.2007.00107.

[48] Haddam M, Chimi H and Amine A. Formulation d'une huile d'olive de bonne qualité. Oilseeds and fats, Crops and Lipids. 2014; 21(5): 2-10.

[49] Bendi Djelloul MC, Amrani SM, Rovellini P, Chenoune R. Phenolic compounds and fatty acids content of some West Algerian olive oils. Communicata Scientiae. 2020, 1: 1-11. DOI : 10.14295/cs.v11i0.3247

[50] Bonanome A, Pagnan A, Biffanti S. Effect of Dietary Monounsaturated and Polyunsaturated Fatty Acids on the Susceptibility of Plasma Low-Density Lipoproteins to Oxidative Modification. Arterioscler Thromb. 1992; 12: 529-533. DOI :10.1161/01.ATV.12.4.529

[51] Lanza B and Ninfali P. Antioxidants in Extra Virgin Olive Oil and Table Olives: Connections between Agriculture and Processing for Health Choices. Antioxidants. 2020; 9 (41): 1-17. DOI:10.3390/antiox9010041

[52] Rocha J, Borges N and Pinho O. Table olives and health. Journal of Nutritional Science. 2020; 9(e57): 1-16. doi:10.1017/jns.2020.50

[53] Castellano JM and Perona JS. Effects of virgin olive oil phenolic compounds on health: solid evidence or just another fiasco? Grasas y Aceites. 2021; 72(2): e404. DOI: 10.3989/gya.0217201

[54] Aridi YS, Walker JL, Wright ORL. The association between the Mediterranean dietary pattern and cognitive health: A systematic review. Nutrients. 2017; 9 (674): 1-23. DOI. org/10.3390/nu9070674.

[55] Nilsson M. Effects of the Mediterranean diet on brain function: Underlying mechanisms [Bachelor Degree Project]. University of Skovde; 2019.

[56] Tripodi P, Schiavi M, Lo Scalzo R. Multi-Scale Evaluation on Two Locations and Digital Fruit Imaging Highlight Morpho-Agronomic Performances and Antioxidant Properties in Chili Pepper Hybrids. Agronomy. 2021; 11(805):1-12. DOI:10.3390/agronomy11040805

[57] Saha S, Walia S, Kundu A, Kaur C, Singh J and Sisodia R. Capsaicinoids, Tocopherol, and Sterols Content in Chili (Capsicum sp.) by Gas Chromatographic-Mass Spectrometric Determination. International Journal of Food Properties. 2015; 18(7): 1535-1545. DOI: 10.1080/10942912.2013.833222

[58] Motukuri K and Jaswanthi N. Chapter 1-9: Hot Pepper (*Capsicum annuum* L.): An Alternative Food to Reduce Micronutrient Deficiencies in Human. In book Capsicum. 2020. DOI:10.5772/intechopen.92198

[59] Bonaccio M, Di Castelnuovo A, Costanzo S, Ruggiero E, De Curtis A, Persichillo M, Tabolacci C, Facchiano F, Cerletti C, Donati MB, De Gaetano G, Iacoviello L. Chili Pepper Consumption and Mortality in Italian Adults, J Am Coll Cardiol. 2019, 74 (25): 3139-3149. DOI: 10.1016/j.jacc.2019.09.068.

[60] Materska M and Perucka I. Antioxidant activity of the main phenolic compounds isolated from hot pepper fruit (*Capsicum annuum* L.). Journal of Agriculture and Food Chemistry. 2005; 53(5): 1750-1756. DOI:10.1021/jf035331k

[61] Wahyuni Y, Ballester AR, Sudarmonowati E, Bino RJ and Bovy AG. Secondary Metabolites of Capsicum Species and Their Importance in the Human Diet. J. of Natural Products. 2013; 76 (4):783-793. DOI:10.1021/np300898z

[62] Chouaibi M, Rezigc L, Hamdic S, Ferrari G. Chemical characteristics and compositions of red pepper seed oils extracted by different methods, Industrial Crops & Products. 2019; 128: 363-370.

[63] Meckelmann SW, Riegel DW, van Zonneveld M, RíosL, Peña K, Mueller-Seitz E and Petz M. Capsaicinoids, flavonoids, tocopherols, antioxidant capacity and color attributes in 23 native Peruvian chili peppers (Capsicum spp.) grown in three different locations. European Food Research and Technology. 2014; 240 (2): 273-283.DOI: 10.1007/s00217-014-2325-6

[64] Olatunji TL and Afolayan AJ. The suitability of Chili pepper (*Capsicum annuum* L.) for alleviating human micronutrient dietary deficiencies. Food Sci Nutr. 2018; 6: 2239-2251. Doi:10.1002/fsn3.790

[65] Ben Mansour-Gueddes S, Tarchoun N, Teixeira Da Silva JA and Saguem S. Agronomic and Chemical Evaluation of Seven Hot Pepper (*Capsicum annuum* L.) Populations Grown in an Open Field. Fruit, Vegetable and Cereal Science and Biotechnology, Global Science Books. 2010; 4(1): 93-97.

[66] Antonious GF. Chapter 2: Capsaicinoids and Vitamins in Hot Pepper and Their Role in Disease Therapy. In: Handbook of Capsaicin and its Human Therapeutic Development. Kentucky State University, Frankfort, KY, USA, 2018; 13-40. DOI.org/10.5772/intechopen.78243.

[67] Cvetkovi CT, Ranilovi CJ, Gajari D, Tomic-Obrdalj H, Šubari CD, Moslavac T, Cikoš AM and Joki CS. Podravka and Slavonka Varieties of Pepper Seeds (*Capsicum annuum* L.) as a New Source of Highly Nutritional Edible Oil. Foods. 2020; 9 (1262): 1-21. DOI:10.3390/foods9091262

13

Importance of the Natural Incidence of the *Fusarium* Genus in Food Crops Established in Northern México

Julio César Ríos Saucedo, María Gabriela Ramírez-Valadez, Saúl Santana Espinoza, Maihualy Martínez-Fernández and Rigoberto Rosales-Serna

Abstract

The incidence of the *Fusarium* genus causing root rot is reviewed in crops showing high importance for food supply and to obtain regular income by farmers in the highlands of Northern México. Pathogen incidence was evaluated under field conditions in multiple sampling locations for common beans (*Phaseolus vulgaris* L.) and several chili peppers (*Capsicum annuum*) local cultivars (landraces and bred cultivars). Five commercial plots for registered and certified seed were also evalu-ated in common beans to be used in the 'seed refreshing program' implemented for the cultivar Pinto Saltillo, considered as the main variety sown in the highlands of México. High *Fusarium* genus incidence and its interactions with other fungi species, such as *Rhizoctonia solani* and *Pythium* spp., cause high losses in plant population, commercial yield and seed quality in food crops grown in Northern México. The natural incidence of plant disease caused by the *Fusarium* genus and its negative effect on crop survival and the reduction of commercial yield and seed quality is fully reviewed. Plant disease resistance, crop breeding and the influence of the environmental conditions were also considered.

Keywords: *Phaseolus vulgaris*, *Capsicum annuum*, root rot, commercial yield, seedborne pathogens

1. Introduction

The *Fusarium* genus includes several fungi species considered as the most important soil and seedborne pathogens causing plant diseases with high yield, quality and economic impact in several food and cash crops sown in the highlands of México [1]. The main effect of the *Fusarium* genus is the root and crown rot observed in vascular bundles of common bean (*Phaseolus vulgaris* L.), chili pepper (*Capsicum annuum* L.), maize (*Zea mays* L.) and other crops used as food, fodder and to obtain cash for urgent expenditures by the poor farmers in Northern México [2]. The main *Fusarium* symptoms in the plants are observed at the soil surface or below the ground including damage in root crown and embryonic root, losses of root water-absorbing surface, leaf wilting and plant death [3, 4]. As a result, low plant densities are commonly observed in common bean and chili commercial plantings in most of the producing areas of Northern México [5, 6].

Under rainfed conditions losses in plant population are aggravated in years registering drought stress [7] after the common bean seedlings emergence. Plant losses are also observed under irrigation [8] when high temperatures and fast drying of the soil surface are registered, mainly after plant emergence (V_1) and the primary leaves unfolded (V_2) stages in common beans [9] and the second week after chili seedlings transplant in cultivated soil. Under those conditions, no post-embryonic adventitious root growth and plant death are observed due to severe damage in the embryonic root caused by the fungi complex (*Fusarium* spp., *Rhizoctonia* spp., and *Pythium* spp.) [10].

Plant losses in the early stages of growth in common bean and chili are considered one of the main causes of reducing yield and the quality of the commercial product (seed and chili pepper: fresh, dried, or processed fruits) [5, 11–13]. Decrement of the income is also observed by farmers, considering that low-input agriculture is the best option for reducing risks in crop production thus obtaining low yields. The actual measurement of the incidence and severity levels of plant pathogens causing root and crown rot, as well as diseases in the aerial plant organs is necessary to establish control strategies according to agroecological and sustainable agriculture. Identification of plant disease agents is the key to the development of effective control and management strategies [14]. Actual evaluation of disease problems related to the *Fusarium* genus in the state of Durango was included in this study regarding the importance of common bean and chili peppers as important food-producing crops.

2. Common bean

Common bean is an important cash crop used as a food, nitrogen-fixing plant and organic matter source [15], thus helping to reach sustainability in the agriculture performed under drought-prone and irrigated areas in the Mexican highlands of Northern México, including the states of Zacatecas, Durango and Chihuahua. In this area near 1.6 million of hectares are annually sown and more than 1.0 million of tons of grain are produced, at a rate of 0.69 t ha^{-1} [16]. Common bean cultivars Pinto Saltillo (improved) and Negro San Luis (landrace) are the most important varieties, according to the planted area and grain volume produced. Sowing recom-mendations include 35 to 50 kg ha^{-1} of seed [17] to obtain plant densities ranging from 100,000 to 120,000 plants ha^{-1}. In despite of recommendations low plant densities (<80,000 plants ha^{-1}) are very common in commercial plantings through the common bean-producing areas of Northern México, thus lower seed yield is also observed across locations and years [2].

Reduction in plant densities reached values lower than 20% of the recommended levels, and the lowest levels (12,500 plants ha^{-1}) were related to the lowest seed yield [5]. Plant densities reduction in early stages of crop growth were related to the low seeding rates, mechanical soil impedance and seedlings death caused by the fungi complex including *Fusarium solani* f. sp. *phaseoli*. Identification and predominance studies of pathogenic fungi are important to establish control strate-gies to reduce disease problems. Studies on fungi genetic and pathogenic diversity are also important for efficient disease control in several crop-producing areas of Northern México. Several strategies were used for fungal disease control including crop genetic breeding, crop rotation and other modern agroecological and chemical management technologies.

2.1 *Fusarium* natural incidence

Twenty root samples were taken during the 2007 growth cycle in 15 common bean commercial plots distributed across "Los Llanos" de Durango and the Valleys of Poanas, Guadiana and Canatlán [10]. Fungi genus predominance was recorded using presence or absence in each root sample. Seventy-five fungi strains were isolated mainly belonging to three genera as follows: 54% for *Fusarium* sp., 10% *Macrophomina* and 2% *Pythium*. *Fusarium* resulted in a genus widely distributed, present in all the sampled locations and within genus three species were identified as *Fusarium oxysporum*, *F. solani* and *Fusarium graminearum* [10].

The predominance of *Fusarium* was observed in 93% of sampled locations and in 75% of the samples was the only genus of fungus found, affecting all the seed commercial classes (pinto, shiny black and flor de junio), while *Macrophomina* was associated mainly with pinto cultivars [10]. Results corroborated previous findings which demonstrated a close relationship between the *Fusarium* genus and common bean root rot in the state of Durango [1]. Systematic studies for fungi phenotypic, pathogenic, and genetic variation need to be implemented to select efficient, sustainable and agroecological disease control methods.

Another study was performed in Durango during 2020, including seven commercial plots at two important common bean producing areas (**Table 1**). The importance of *Fusarium* fungus was corroborated and variation in presence values from 50 to 100% were obtained. Other fungi genera were found at different levels of presence including *Rhizopus* (0–40%), *Pythium* (10–30%), *Erysiphe* (10–20%) and *Cercospora* (10–20%). The *Fusarium* fungus remains as the major plant disease problem in Durango, where integral control programs are necessary including crop genetic improvement, crop rotation, and other agroecological control methods.

Preventive chemical control and qualified (registered and certified) seed use are also considered. *Fusarium* control is an important issue due to some fungal species on stored common bean can also release mycotoxins (fumonisin) which causes human mycotoxicoses upon consumption, esophageal cancer and interferes with sphingolipid metabolism [18].

		Fungi genera				
Municipality	**Plot**	***Fusarium***	***Rhizopus***	***Pythium***	***Erysiphe***	***Cercospora***
Durango	1	50	40	10	20	20
	2	100	40	30	10	20
	3	80	40	20	10	20
[1]G. Victoria	[2]J. G. R.	60	30	20	10	20
	A. A.	70	30	20	20	10
	I. A.	60	0	30	10	20
	C. C.	70	0	10	10	10

[1]*G. Victoria = Guadalupe Victoria.*
[2]*J. G. R. = José Guadalupe Rodríguez, A. A. = Antonio Amaro, I. A. = Ignacio Allende and C. C. = Calixto Contreras.*

Table 1.
Isolation frequency of pathogen fungi related to common bean root rot at different municipalities of Durango, México. 2020.

2.2 Genetic improvement

Few systematic studies have been implemented in Durango for root diseases, mainly due to difficulties observed for plant root extraction, destructive sampling methods, and laboratory requirements for pathogen isolation and conservation, as well as for selecting crop-resistant germplasm. Studies concluded that low genetic diversity was observed for resistance in the plant pathogen-host represented by multiple common bean cultivars belonging to different genetic races and variable seed sizes and color. Sources of resistance against *Fusarium* wilt were identified in flor de mayo (pink) germplasm [19], mainly related to the Jalisco Race [20].

Low genetic resistance to *Fusarium* spp. has been identified in cultivated common bean populations, and the genetic resistance is quantitative [21] and hence, strongly influenced by the environment [22]. Other studies reported dominance in the control of the character with additive effects in common bean indicating that selection should be easy using efficient inoculation and selection methods [23]. In Durango, no direct selection was performed for *Fusarium* root rot in common bean, however, results indicated that resistance to FSP was more frequent in black beans [11]. In despite of the general observations, pinto seeded cultivars (Durango Race) showed capability for rapid adventitious root growth to maintain water absorption from the superficial soil layers and reduce losses in plant population.

2.3 Crop rotation

A common bean monocropping system is a common agricultural practice in most of the production areas in Northern México, thus aggravating problems and damage caused by *Fusarium* and other soilborne and seed transmitted pathogens. Crop rotations under rainfed conditions depend on the rain occurs during the May to August period. Early rains (May and early June) favor maize plantings while oats sowing is preferred when late raining periods (after middle August) are observed. Common bean is preferred to be sown when the rains are registered in late June to middle July. Under irrigation forage crops such as corn, sorghum, grasses, lucerne and oats are preferred due to pressure exerted by cattle farmers. Systematic crop rotations are required in Durango to reduce plant root and aerial pathogen problems in several crops and advances in sustainability could be also achieved by reducing water use by planting low water requirement crops.

2.4 Agroecological control methods

Trichoderma sp. showed high efficiency as a control agent for a wide range of aerial and soilborne plant pathogens and this trait makes it an excellent candidate for controlling saprophytic growth of *Fusarium* [24]. Some attempts were made to evaluate to control efficiency of this natural soil organism by reducing plant damages and yield losses caused by *Fusarium*. Results are considered ambiguous and *Trichoderma* use are not yet included in common bean crop management recommendations. Other options have been explored such as plant-based biopesticides [25] without actual use at the commercial level.

2.5 Chemical control

Disease chemical control starts with the seed treatment, but in most of the production areas in Durango the use of a fungicide is considered only in qualified seed production programs, which include several quality categories: basic, registered, certified and declared seeds [26, 27]. Most of the farmers consider chemical treatment as a "fallacy" used only to justify the increment in seed prices without additional benefits for plant disease control under field conditions. Chemical products recommended in common bean seed treatment are: Metacaptan® (Captán + Metoxichloro), Thiram (Tebuconazole + Thiram), Terrazán [(Quintozeno-(pentacloronitrobenceno)], vitavax (Carboxín + Captán), Ridomil (Metalaxil-M) and Benlate (Benomyl).

The field visual evaluations showed some beneficial effects of the chemical seed treatment for increment the seedling emergence and plant survival at the early stages of the common bean development. Although systematic evaluation of subsequent fungicide effects on seedling and adult plant disease control and seed yield are necessary to reinforce the recommendations for its use. Several fungicides showing contact and systemic effects are recommended to reduce root rot incidence and severity, but results are not conclusive. Plant genetic resistance is preferred across the common bean production areas along México where several cultivars have been developed [28].

2.6 Qualified seed

Qualified seeds used in México for common bean commercial plantings are known as basic (foundation), registered (first generation certified), certified (second generation certified) and "enabled" seed. Standard seed treatment include fungicide (Metacaptan®, Thiram, Terrazán, Vitavax, Ridomil and Benlate), insecticide (Deltamethrin) and rhodamine as a colorant [26]. Effectivity on seed treatment needs to be evaluated due to the increase in the number of companies dedicated to seed production under several environmental conditions and variation in compliance with the regulations. In 2021, five seed lots from different sources were evaluated considering quality, which includes the genetic, physiological, physic, and sanitary traits (**Table 2**). The standard germination test is the most used probe to evaluate the physiological quality of a seed lot.

The pathogen attack on seeds is one of the factors that leads to the physiological quality loss, reducing the germination rate and the vigor of the seed lots, which end up precluding the final stand of the crop, resulting in productivity and economic losses to the farmer. Fungi are considered as the most important among pathogens due to the higher number of species and the damage caused both in yield and seed quality [29]. The mixture containing Carbendazim + Thiram in its composition are efficient in the control of pathogens regardless of the application time of products [29].

The sanitary quality of the seed used in México needs to be evaluated to reduce problems during the seed germination and seedling emergence periods, mainly under field conditions. Some chemical products were identified for their efficient control of different fungus although some phytotoxicity effects were also observed and delayed protection by fungicide controlling *Fusarium* on seeds [29]. In 2021, five

Seed Lot (origin)	Category	Year of production	Reception date	Amount (kg)
1. Chihuahua 1	Registered	*SS-2020	04/12/20	3000
2. Chihuahua 2	Registered	SS-2020	15/12/20	3000
3. Durango 1	Certified	SS-2016	15/02/2021	10,030
4. Sinaloa	Registered	AW-20	30/03/2021	3000
5. Durango 2	Registered	SS-2020	15/06/2021	9000

**SS = spring–summer growth cycle; AW = autumn-winter growth cycle.*

Table 2.
Reception data of five qualified seed lots to be used in reinforcing Pinto Saltillo common bean commercial plantings in Durango, México.

seed commercial lots (**Table 2**) were evaluated according to the physiological quality tests at the INIFAP´s Valle del Guadiana Experiment Station, located in Durango, Méx. Standard germination test was performed, using a soil-based substrate, and considering the SNICS (Servicio Nacional de Inspección y Certificación de Semillas) recommendations for sample size (30 samples) in each seed lot. The seed lot number 4 from Sinaloa registered the highest emergence level (96%), reaching that value in the shortest period (8 days after planting; DAP); while, lot number 3 (from Durango) showed the lowest emergence value (65%), reached at 12 DAP (**Figure 1**).

Reductions in seed germination levels were related to the production environment and storage conditions and duration. Other negative factors were mechanical damages caused during the harvest-threshing and seed cleaning processes, as well as the seed fungi load and chemicals used for the seed treatment [29, 30]. Seedlings wilting was also observed during the germination test due to soil infestation and seed contamination by fungi. Therefore, seed studies for soil and seedborne pathogen load were performed under controlled laboratory conditions.

2.7 Seed health tests

Pinto Saltillo is an improved common bean cultivar showing high yield and disease susceptibility in aerial (Common Bacterial Blight) and root zone (*Rhizoctonia* spp., *Fusarium* spp., and *Pithyium* spp.). Most of these diseases are seedborne and the infestation/contamination of the seed may occur during harvest-threshing activities, processing and handling. The pathogen may, thus, be carried with the seeds in three ways: Admixture (pathogen are independent of seed but accompany them), External (pathogen present in seed surface as spores, oospores and chlamydospores) and internal (pathogens establish within the seed with the definite relationship with seed parts) [31]. The pathogen *Fusarium* is soil as well as seedborne in nature and the colonization percent of *F. solani* was highest as compared to other isolated fungi. Seed germination rate was also reduced (50%) in soil infested with *F. solani* where seedlings mortality reached 93.3% [32].

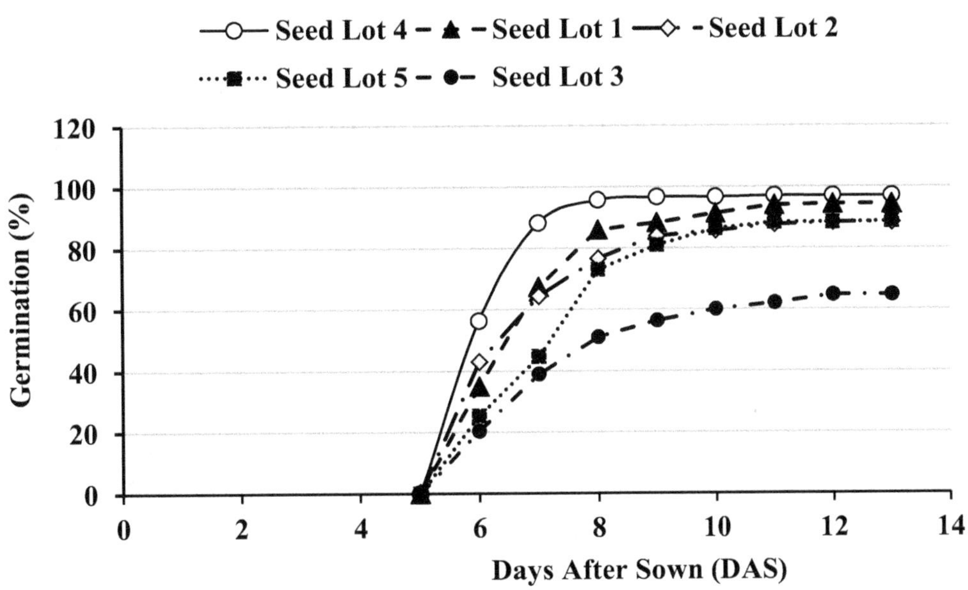

Figure 1.
Germination standard test performed in five qualified common bean seed lots obtained in México. 1) Chihuahua 1, 2) Chihuahua 2, 3) Durango 1, 4) Sinaloa y 5) Durango 2.

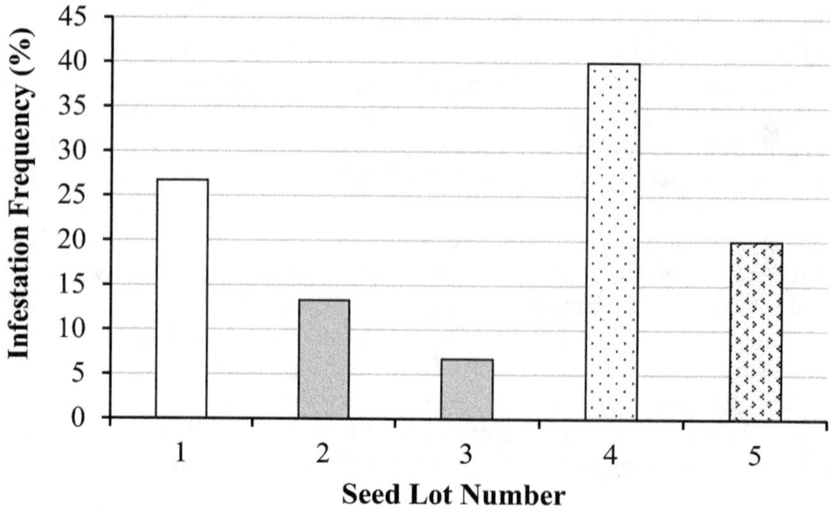

Figure 2.
Fusarium fungus infestation frequency observed in seed lots from different states to be used in Durango. 1) Chihuahua 1, 2) Chihuahua 2, 3) Durango 1, 4) Sinaloa y 5) Durango 2.

Agar plate is considered the most common method used for identification the of seedborne fungi. In 2020, *Fusarium* presence in Pinto Saltillo seed lots from different origins (**Table 1**) was determined by triplicate placing seeds onto sterile agar media (Potato Dextrose Agar: PDA) to encourage the growth of the fungus [33]. Variation of the infestation frequencies was registered among seed lots from different seed sources from México (**Figure 2**). High frequencies (40.0%) were detected in Pinto Saltillo seed lots produced in Sinaloa during the Autumn-Winter (2020–2021) growth cycle where several *Fusarium* hospedant were cultivated (chickpea, tomato, common beans and maize) [2].

Fusarium isolates from Sinaloa, Méx., showed differences in aggressiveness; and *F. falciforme* was the most aggressive compared to *F. oxysporum* [34], and isolates of both complexes triggered similar aerial symptoms of yellowing and darkening of the vascular tissues in tomato plants. But only *F. falciforme* isolate triggered necrosis in the plant crowns [34]. Seed lot 3 showed low frequencies for *Fusarium* fungus incidence (6.7%), mainly due to the longest storage period since was produced in the 2016 Spring–Summer growth period. Long storage period reduced pathogen fungus load but low seed germination and seedlings damage (injured leaves) were also observed.

3. Chili peppers

In México, generic name of chili pepper (chile; *C. annuum* L.) is used to denominate several plant cultivars mainly known with a local names such as: chile ancho, jalapeño (processed chipotle), serrano, mirasol (dried guajillo), and pimiento morrón reaching the 70 and 80% of the national production [35]. Chili pepper is one of the most important vegetable crops used in México as condiment and food flavor and its also considered as an important cash crop [36] grown in several production areas in Northern México, providing additional income to the farmers.

In the Northern highlands of México, the planted area for fresh fruit harvest of chili pepper in 2020 reached 37,440 ha in Zacatecas, 30,772 ha in Chihuahua and 4,136 ha in Durango [16]. In Durango, several chili cultivars and landraces are planted [jalapeño, poblano-ancho, puya, mirasol-guajillo, árbol, cola de rata and tornachile (chile güerito)]; while in Chihuahua and Zacatecas, the jalapeño, serrano and habanero cultivars are preferred. The highest chili production is obtained in Chihuahua (722,937 t) where the yield rate is 24.0 t ha^{-1}. The state of Zacatecas produces 458,943 t and the yield average is 12.4 t ha^{-1}. In the state of Durango the chili production overpasses 48,035 t and the lowest yield at the North Central region of the Mexican highlands is obtained (11.6 t ha^{-1}) [16].

Chili crop management system includes seed obtention from dried fruits, sowing and nursery growth (almácigo), and transplant under field conditions, using rows 0.81 to 1.20 m apart [35]. Modern management techniques include the use of mulch and drip irrigation [37] to increment yield and water productivity. Several plant pathogens are observed in chili plantations established in Durango, causing phytosanitary problems, low yield and reduced fruit quality; as well as generalized plant or whole plot losses. Plant pathogen problems include viruses (Cucumber Mosaic: CMV, Potato Y: PYV, Alfalfa Mosaic: AMV, Tobacco Mosaic: TMV and TEV), fungi (*Phytophthora capsici*, *Rhizoctonia solani*, *Fusarium* spp., *Pythium* spp.) [12] and bacteria (*Xanthomonas campestris* pv. *vesicatoria* and *Pseudomonas syringae* pv. *tomato*) [38].

The pathogenic syndrome known as chili plant wilting 'Secadera' (CPW), is the most important disease in all the producing areas of México, causing total yield losses (100%) and the planting area was reduced by 60% in some states [39]. CPW is mainly caused by the obstruction of the vascular bundles provoked by the phytopathogen fungi infection (*Fusarium oxysporum, Rhizoctonia solani*, and *Phytophthora capsici*) in roots or root-crown [34]. The plant pathogen, mainly *Fusarium*, relative importance need to be evaluated due to its alimentary, health, economic and social implications.

3.1 *Fusarium* natural incidence

Root and crown samples were taken when 'Secadera' (Damping-off and CPW) symptoms (symptomatic plants) were observed at the main chili-producing areas in the state of Durango.

3.1.1 Study 1

Eleven municipalities were included (**Table 3**) and random sampling included 26 commercial plots established under irrigation in each of the chili pepper production areas across the state of Durango where temperate and warm climates are registered. Plots were georreferentiated for map construction (**Table 3**). Direct sampling was made in plots showing typical Damping-Off and CPW symptoms, such as: yellowing and wilting in upper leaves, wilting symptoms in all parts of the plant; leaves showing dark-green color remaining attached to the plant, root crown narrowing and plant death. Samples consisted of 10 plants in each plot, which were dissected and tissue samples were taken in the root crown and embryonic root. Samples were transported in paper bags with an identification label, including municipality, location, geographic coordinates and crop cultivar, then were sun-dried and stored at room temperature until processing.

3.1.2 Study 2

Seven chili pepper sampling sites were included at two municipalities in the state of Durango (**Table 4**), plots were mainly established under irrigation in the temperate climate regions. Plots were georreferentiated for map construction (**Table 4**). Direct sampling was made in plots showing typical Damping-Off and CPW symptoms. Samples consisted of 10 plants in each plot which were dissected,

Sample number	Municipality	Plot	Geographic localization	Cultivar local name
1	Simón Bolívar	Simón Bolívar	N 24° 45′ 42.0″; W 103° 10′ 51.5"	Ancho
2	Simón Bolívar	Flores Magón	N 24° 44′ 43″; W 103° 10′ 29.0"	Ancho
3	Lerdo	El Refugio	N 25° 25′ 59.0″; W 103° 44′ 57.3"	Guajillo
4	Lerdo	21 de marzo	N 25° 26′ 45.2″; W 103° 45′ 13.0"	Guajillo
5	Mapimí	Perimex	N 26° 32′ 36.6″; W 104° 08′ 15.8"	Jalapeño
6	Nazas	Francisco Sarabia	N 25° 18′ 12.6″; W 103° 58′ 20.5"	Guajillo
7	Nazas	Agustín Melgar	N 25° 16′ 25.3″; W 103° 58′ 20.5"	Guajillo
8	Nazas	Lázaro Cárdenas	N 25° 16′ 25.0″; W 104° 02′ 43.8"	Guajillo
9	Nombre de Dios	Francisco Munguía	N 23° 47′ 33.6″; W 104° 06′ 15.2"	Puya y Árbol
10	Peñón Blanco	J. Agustín Castro	N 24° 38′ 51.0″; W 103° 56′ 01.1"	Ancho
11	Poanas	Dago 1	N 23° 58′50.2″; W 104° 03′ 54.0"	Mirasol
12	Poanas	Pozo 12	N 23° 59′ 29.3″; W 104° 04′ 00.7"	Ancho y Puya
13	Poanas	UTP 1	N 23° 56′ 26.7″; W 104° 02′ 56.8"	Ancho y Puya
14	Poanas	UTP 2	N 23° 55′ 53.9″; W 104° 02′ 57.8"	Ancho y Árbol
15	Poanas	UTP 3	N 23° 56′ 26.7″; W 104° 02′ 56.8"	Ancho
16	Poanas	Pilares	N 23° 52′ 24.7″; W 104° 02′ 40.2"	Puya
17	Rodeo	El Parián	N 25° 09′ 45.2″; W 104° 32′ 40.2"	Ancho
18	Rodeo	Primo de Verdad	N 24° 54′ 52.7″; W 104° 28′ 5.5"	Ancho y Cola de Rata
19	Rodeo	La Cuesta	N 25° 01′ 47.7″; W 104° 28′ 40.9"	Ancho
20	S. Juan del Río	El Crucero 1	N 24° 46′ 58.4″; W 104° 30′ 56.9"	Ancho
21	S. Juan del Río	El Crucero 2	N 24° 46′ 51.3″; W 104° 31′ 14.8"	Ancho
22	S. Juan del Río	Francisco de Ibarra	N 24° 47′ 26.3″; W 104° 30′ 38.2"	Ancho y Tornachile
23	S. Juan del Río	Santa Rosalía	N 24° 53′ 44.6″; W 104° 26′ 17.4"	Ancho y Cola de Rata
24	S. Juan del Río	José María Patoni	N 24° 52′ 51.7″; W 104° 26′ 17.1"	Ancho

Sample number	Municipality	Plot	Geographic localization	Cultivar local name
25	S. Pedro del Gallo	La Laborcita	N 25° 29′ 21.3″; W 104° 24′ 03.0”	Ancho y Guajillo
26	S. Luis del Cordero	San Luis del Cordero	N 25° 24′ 42.0″; W 104° 16′ 03.0”	Árbol

Table 3.
Geographic localization of plots included in chili sampling implemented in plants showing plant wilting (CPW) symptoms at different municipalities of the state of Durango, México.

Sample number	Municipality	Plot	Geographic localization	Cultivar local name
1	Durango	CEVAG 1	N 24° 44′ 43″; W 103° 10′ 29.0”	Ancho
2	Durango	CEVAG 2	N 24° 44′ 43″; W 103° 10′ 29.0”	Ancho
3	Durango	CEVAG 3	N 24° 44′ 43″; W 103° 10′ 29.0”	Ancho
4	G. Victoria	J. G. Rodríguez	N 25° 26′ 45.2″; W 103° 44′ 57.3”	Guajillo
5	G. Victoria	A. Amaro	N 26° 32′ 36.6″; W 104° 08′ 15.8”	Jalapeño
6	G. Victoria	I. Allende	N 25° 18′ 12.6″; W 103° 58′ 20.5”	Guajillo
7	G. Victoria	C. Contreras	N 25° 16′ 25.3″; W 103° 58′ 20.5”	Guajillo

Table 4.
Geographic localization of plots included in chili sampling implemented in plants showing plant wilting (CPW) symptoms at two municipalities of the state of Durango, México. 2020.

and tissue samples were taken in the root crown and embryonic root. Samples were transported in paper bags with an identification label, including municipality, location, geographic coordinates and crop cultivar, then were sun-dried and stored at room temperature until processing.

3.2 Morphological characterization

Two classes of fungi (Anamorphic and Oomycota) were isolated including three different genera morphologically differentiated (**Table 5**). The most abundant genus was *Fusarium*, followed by *Rhizoctonia* and in a lower extent the Omicete *Pythium*. *Fusarium* were detected in 100% of the samples collected in 8 locations at municipalities of Simón Bolívar (2), Lerdo (1), Rodeo (1), San Juan del Río (3) and San Pedro del Gallo (1). This plant pathogen was present in variable proportions (18–100%) of samples at 25 locations (96%). *Rhizoctonia* were detected in 100% of the samples at five locations, such as: Simón Bolívar (2), Nombre de Dios (1), Poanas (1) and San Luis del Cordero (1). This fungus was present in 23 (88.5%) of the total samples with presence levels ranging from 18 to 100% across locations. A low incidence level was observed for *Pythium* and absence was registered at 16 locations (61.5%), with presence levels from 20 to 70% of the samples with the highest value (70%) at the location of J. Agustín Castro.

Results of the isolation frequency showed that *Fusarium* was the fungus present in most of the samples of plant material showing CPW symptoms, regardless of the sample site of origin, except for San Luis del Cordero. *Fusarium* was the widespread and frequently pathogenic genus of plant fungus, followed by *Rhizoctonia*, while in most of the sampling sites absence of oomycetes (*Pythium*) was found, mainly at the

	Municipality	Plot	Frequency (%)		
			Fusarium sp.	*Rhizoctonia* sp.	*Pythium* sp.
1	Simón Bolívar	Simón Bolívar	100	100	44
2	Simón Bolívar	R. Flores Magón	100	100	45
3	Lerdo	El Refugio	100	0	0
4	Lerdo	21 de Marzo	60	30	20
5	Mapimí	Perimex	60	40	60
6	Nazas	Francisco Sarabia	40	30	60
7	Nazas	Agustín Melgar	80	50	30
8	Nazas	Lázaro Cárdenas	40	70	50
9	Nombre de Dios	Francisco Munguía	40	100	0
10	Peñón Blanco	J. Agustín Castro	80	60	70
11	Poanas	Dago 1	33	67	0
12	Poanas	Pozo 12	73	27	0
13	Poanas	UTP 1	50	70	0
14	Poanas	UTP 2	89	33	0
15	Poanas	UTP 3	63	88	0
16	Poanas	Pilares	18	100	0
17	Rodeo	El Parián	63	38	38
18	Rodeo	F. Primo de Verdad	100	33	0
19	Rodeo	La Cuesta	75	58	0
20	San Juan del Río	El Crucero 1	100	0	0
21	San Juan del Río	El Crucero 2	100	18	0
22	San Juan del Río	Francisco de Ibarra	100	0	0
23	San Juan del Río	Santa Rosalía	70	60	0
24	San Juan del Río	José María Patoni	50	50	0
25	San Pedro del Gallo	La Laborcita	100	63	38
26	San Luis del Cordero	San Luis del Cordero	0	100	0

Table 5.
Isolation frequency of pathogen fungi related to chili pepper plant wilting at different municipalities of Durango, México.

municipalities of the semi-desertic region (Lerdo, San Juan del Río and Rodeo), but also in the highland valleys of Poanas and Nombre de Dios. Absence of *Pythium* was related to the phenological stage of sampled chili pepper populations since this fungus effect is mainly observed in the early stages of crop development. Soilborne pathogen *Fusarium oxysporum* and *Rhizoctonia solani* are the most common diseases causing root-rot and plant wilt in chili pepper cropping fields [36, 40]. These fungi are also observed in common bean and cereals planted in Durango for fodder and food, where the contamination with *Fusarium* species is one of the major sources of mycotoxins [41].

In the municipalities belonging to 'La Laguna' region (Mapimí and Nazas), the frequency of isolation of *Fusarium* was like that observed in other sampling sites, however, that of *Rhizoctonia* decreased and *Pythium* presence increased considerably compared to the municipalities of the highlands region (Poanas) and that of the semi-arid sampling sites (**Table 5**). The use of mulch influenced increments for *Pythium* presence at some locations such as J. Agustín Castro.

In the second study, *Fusarium* also showed the highest presence scores (50 to 90%) at "Valle del Guadiana" and "Los Llanos" regions (**Table 6**). Other fungi species related to the common bean cropping systems such as *Uromyces* and *Pythium*, also showed high presence levels. Other cosmopolitan fungi species *Rhizopus* (Zygomycetes) registered a high presence in plant samples, due to its omnipresent nature as an air contaminant, fast growing rate, and versatility of growth conditions (temperature and relative humidity) [42].

The low presence of *Phytophthora* fungus was observed in most of sampling sites (85.7%), due to increased novel sowing areas for chili peppers opened under irrigation at the municipalities of Durango and Guadalupe Victoria. Increments in the chili pepper area were related to recurrent crop complete losses registered in the main producing area of Poanas, Villa Unión and Nombre de Dios. Potato plantations also influenced the presence of *Phytophthora* fungus in both studied municipali-ties. Other fungi species (*Erysiphe* and *Cercospora*) were found, causing mildew in several crops and leaf or pod spots in common beans. Several fungi genera were detected in cultivated soils of Durango, causing severe economic losses in horticul-tural and agricultural crops.

3.3 Morphological identification

Different fungi and oomycetes show adaptation under different growth media, temperature, and light quality, then producing consistent characteristics that can be used for morphological identification [43]. The colony morphology of the *Fusarium* fungi species of the six most frequent isolates obtained in Durango was determined in pure culture using three different culture standard media: PDA (Potato Dextrose Agar, Difco®), Corn Flour Agar (CA) [44], and SDA (Sabouraud Dextrose Agar, Difco®) also known as Spezieller Nährstoffarmer Agar or Special Low-Nutrient Agar (SNA). Several characteristics were evaluated in fungi colonies, such as: pigmentation (color and hue) of the surface on the front and back of the colony, texture of the colony surface (cottony, resupinate, velvety, powdery, crustaceous, soaked, embedded, yeast-like, sticky, homogeneous or heterogeneous, presence or absence of elevation), margin type of the colony (smooth, regular, irregular, restricted, diffuse), pattern (radiated, flower-shaped or arachnoid), formation of resistance structures (sporodochia), mycelial type and growth rate.

Municipality	**Plot**	*Fusarium*	*Uromyces*	*Rhizopus*	*Phytophtora*	*Pythium*	*Erysiphe*	*Cercospora*
Durango	1	60	20	70	10	0	0	0
	2	70	20	0	20	0	0	0
	3	90	0	50	0	0	0	0
Victoria	*J.G.R	80	30	60	10	10	10	0
	A. A.	70	20	30	20	20	20	10
	I. A.	70	10	40	30	0	0	0
	C. C.	50	0	40	20	0	10	0

**J. G. R. = José Guadalupe Rodríguez, A. A. = Antonio Amaro, I. A. = Ignacio Allende, C. C. Calixto Contreras.*

Table 6.
Isolation frequency of pathogen fungi related to chili pepper root rot at different municipalities of Durango, México.

3.4 Colony characterization of *Fusarium* sp.

Strain EV1r. The growth of *Fusarium* sp. EV1r in PDA medium developed a fast-growing colony (4 days) with beige color (**Figure 3A**). The texture of the colonial surface was cottony presenting a smooth and irregular margin. On the other hand, the development of the strain e EV1r in corn flour agar (CA) developed a fast-growing white colony with a radial pattern and regular margin with a powdery texture (**Figure 3B**). In addition, the EV1r strain was also grown in SNA medium showing rapid growth (5 days) with a white front and back color with the presence of mycelium (**Figure 3C**).

Strain H1Zra. The growth of *Fusarium* sp. strain H1Zra in PDA medium developed a fast-growing colony (5 days) showing white color on both sides (**Figure 4A**). The texture of the colonial surface was cottony with the presence of aerial mycelium and irregular margins. On the other hand, the development of the *Fusarium* H1Zra strain on corn flour agar (CA) developed a fast-growing colony with white color, radial pattern, and regular margin. A velvety texture with the absence of aerial mycelium (**Figure 4B**) was also observed. In SNA (Synthetic Nutrient-Poor Agar) medium, the H1Zra strain showed slow growth (8 days) with an opaque colony and little mycelial development (**Figure 4C**).

Strain H1Zrb. The growth of *Fusarium* sp. H1Zrb in PDA medium developed a fast-growing colony (4 days) of white color with a slight yellow color in the periphery of the colony (**Figure 5A**). The texture of the colonial surface was velvety with the presence of mycelium with a regular margin. The development of the H1Zrb strain on corn flour agar (CA) developed a fast-growing white colony with a radial pattern and regular margin with a velvety and powdery texture with the absence

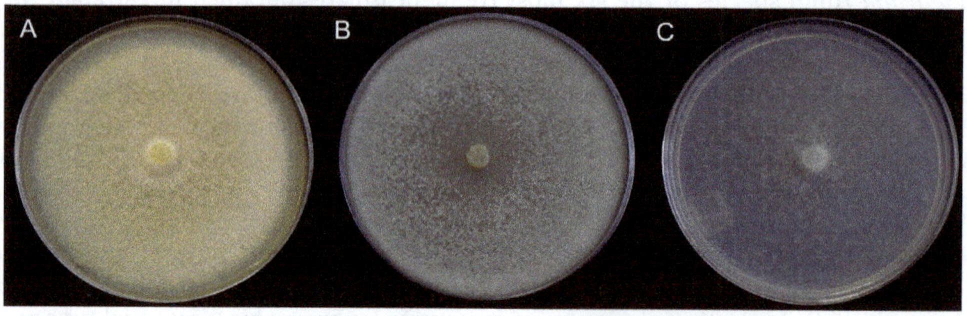

Figure 3.
Colony morphology study in Fusarium *sp. strain EV1r. Growth media: A) potato dextrose agar-PDA, B) corn flour agar-CA, and C) special low-nutrient agar-SNA.*

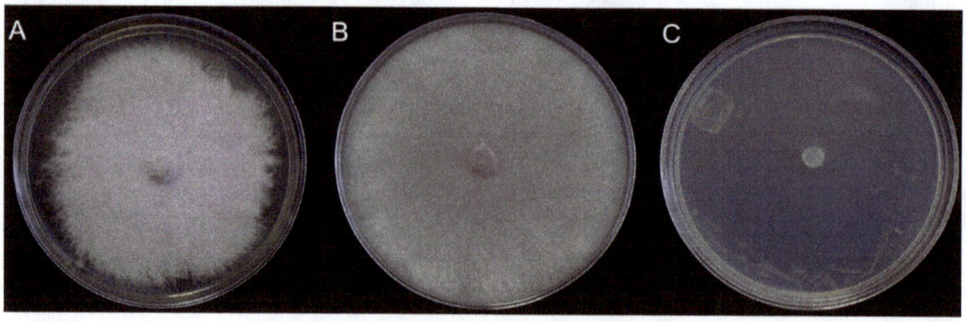

Figure 4.
Colony morphology study in Fusarium *HIZra strain. Growth media: A) PDA B) CA C) SNA.*

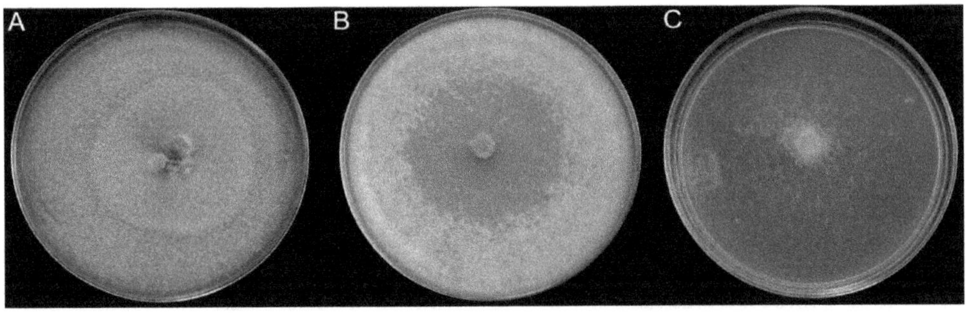

Figure 5.
Colony morphology study in Fusarium sp. strain HIZrb. Growth media: A) PDA B) CA C) SNA.

of aerial mycelium (**Figure 5B**). In SNA medium, the *Fusarium* sp. H1Zrb strain showed rapid growth with an opaque white colony with low mycelium development (**Figure 5C**).

Strain K4Zr. The growth of *Fusarium* sp. K4Zr in PDA medium developed a fast-growing white colony with the presence of growth rings (**Figure 6A**). The texture of the colonial surface is cottony with the presence of aerial mycelium with a regular margin. The development of the K4Zr strain on corn flour agar (CA) developed a rapidly growing white colony with a radial pattern and regular margin with a cottony texture with the presence of floccose mycelium (**Figure 6B**). On the other hand, in the SNA medium, the K4Zr strain showed slow-growth with an opaque white colony with a flat texture with little development of mycelium (**Figure 6C**).

Strain K5Zr. The *Fusarium* sp. strain K5Zr inoculated on PDA medium developed a slow-growth colony (8 days) showing opaque yellow color, with a floral pattern and irregular margin (**Figure 7A**). The texture of the colonial surface was creamy with the absence of aerial mycelium. The development of the K5Zr strain on corn flour agar (CA) developed a rapidly growing white colony with a radial pattern and regular margin with a cottony texture and abundant aerial mycelium (**Figure 7B**). In the SNA medium, the K5Zr strain showed restricted and diffuse growth with an opaque white colony and irregular margin (**Figure 7C**).

Strain C3WC. The growth of *Fusarium* sp. strain C3WC in PDA medium developed a slow-growing colony (4 days) of white color, with regular pattern and margins (**Figure 8A**). The texture of the colonial surface was flat and velvety with the absence of aerial mycelium. On the other hand, the development of the C3WC strain on corn flour agar (CA) developed a colony of white color with a slight brownish color of rapid growth. Radial pattern and regular margin with a cottony

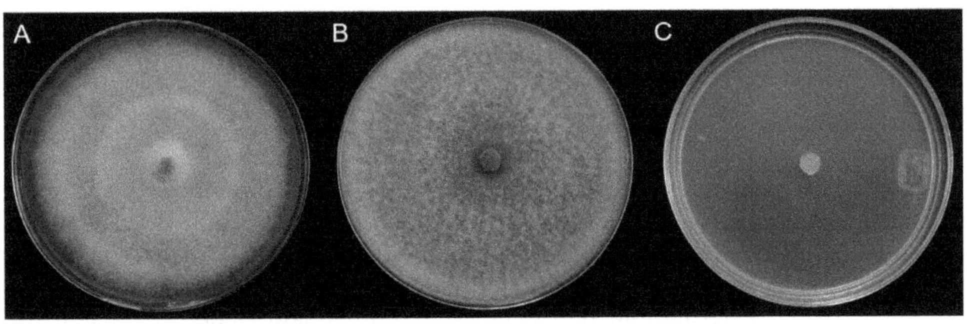

Figure 6.
Colony morphology study in Fusarium sp. strain K4Zr. Growth media: A) PDA B) CA C) SNA.

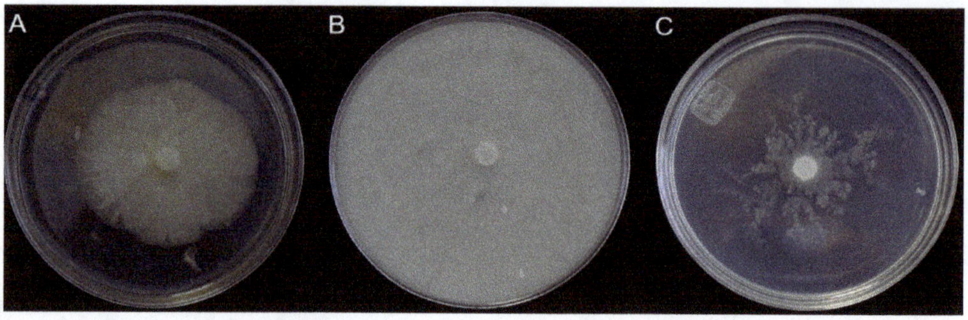

Figure 7.
Colony morphology study in Fusarium sp. strain K5Zr. Growth media: A) PDA B) CA C) SNA.

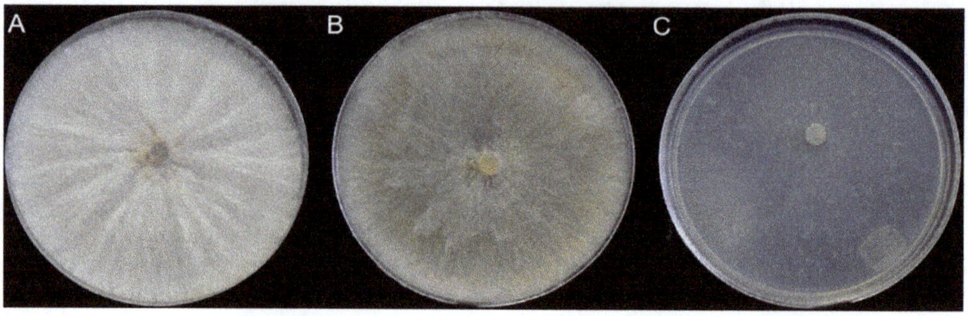

Figure 8.
Colony morphology study in Fusarium sp. strain C3WC. Growth media: A) PDA B) CA C) SNA.

texture and abundant aerial mycelium were also observed (**Figure 8B**). In the SNA medium, the C3WC strain showed a slow-growth colony with an opaque white color and irregular margins and little mycelium development (**Figure 8C**).

3.5 Genetic improvement

In México, 19 accessions of native chili pepper collected in the state of Morelos and 11 serrano chili accessions were selected considering its resistance to *Phytophthora capsici* that can be used in crop breeding [45, 46]. Twenty-six plant accessions were also identified in chili pepper with at least one individual showing resistance to *Fusarium* spp. and only two accessions from the gene bank resulted resistant to *P. capsici* and the mix including *Fusarium, Phythophtora* and *Rhizoctonia* [13]. Despite the germplasm selection for soil and seedborne fungal disease resistance no common bean and chile pepper cultivars for a specific response to soil fungi complex have been released in México.

3.6 Crop rotation

Crop rotation in the chili pepper production areas is influenced by farmer's tradition and the high economic income obtained with the fresh, dried and processed fruits, difficulting changes in the cropping systems. A similar response was observed for the common bean production under rainfed and irrigation monocropping systems. Then agronomic management recommendations need to be adjusted considering agroecological practices, including crop rotation, that contributes to improve the productivity and sustainability of local agroecosystems.

3.7 Agroecological practices

In common bean, chili pepper, maize and most of the crops, agroecological practices and integrated management systems need to be implemented and systematized. Agroecological practices are poorly used in current agriculture in North-Central México, and some components need to be validated at the commercial level including biofertilizers, organic matter incorporation into the soil and the use of natural pesticides, as well as crop choice and crop rotation. Other options include intercropping, relay intercropping, agroforestry with timber, fruit or nut trees; allelopathic plants use (sunflower), direct seeding into living cover crops or mulch, reduced tillage, drip irrigation, biological pest control, and cultivar choice [47].

In Durango, marginal advances were achieved in organic fertilizer production and use, most of the products are not often available in sufficient quantity. Industrial production has been obtained only for compost in "La Laguna" region but high prices have been observed making it unaffordable for farmers. Lombricompost, fulvic acid and other liquid biofertilizers were also produced at low amounts and short time period effects, variable composition, and ambiguous results have been reported. Some interesting results were obtained by using Biological Nitrogen Fixing bacteria (*Rhizobium* spp. and *Azotobacter* spp.) and mycorrhizic fungi, but their production and distribution need to be reinforced. High biomass producing species with appropriate carbon to nitrogen ratio (25) has been selected (*Pennisetum* sp.) to reduce costs for direct organic matter incorporation into soils [48], stabilize pH (6.5–7.0) and naturally release soil minerals for plant nutrition.

Studies on organic pesticides need to be strengthened obtain clear results and to generate recommendations for commercial plantings of common bean, chili pepper and other food and fodder crops. Crop choices have been explored in Durango, using canola (*Brassica* spp.), chickpea (*Cicer ariethinum*), amaranth (*Amaranthus* spp.), barley (*Hordeum vulgare*), sunflower (*Helianthus annuus*), sorghum (*Sorghum bicolor*), and Oats (*Avena sativa*). Some problems need to be solved to improve adoption programs for these crops, such as mechanization from sown to harvest, efficient production storage and commercialization process.

Agroforestry with timber (Scott's Pine: *Pinus greggii*) was implemented using governmental programs, but then abandoned due to prolonged technical periods (10 to 12 years), poor technical support, and food requirement by farmer's families. Studies corroborated that the common bean is better adapted in early years of Scott's pine plantations compared to forage crops (oats and maize) [49]. Increments in Wichita pecan tree (*Carya illinoensis*) plantation area have been observed in the last 5 years in Durango, due to high prices of the pecan nuts and commercial compe-tence to apple production, although influence on local agriculture has not yet been determined. Sunflower is an allelopathic crop which has been used in Durango to reduce problems observed with perennial grasses, mainly bermuda grass (*Cynodon dactylon*). Although, difficulties has been also observed for seed supply and during the crop harvest and seed (achene) commercialization processes.

Direct seeding into living cover crops was probed without success due to pressure exerted by beef/dairy livestock production and the preferential use of crop residues and cover crops as fodder instead as a natural amendment, then low organic matter content is commonly registered in agricultural soils. Low organic matter combined with alkaline reaction in the soils reduces the availability of nutrients, presence of beneficial microorganisms, water infiltration and retention, and then aggravating drought, plant nutrient deficiencies and disease problems in several food and cash crops. Mulch and drip irrigation is used in some cash crops, even common beans, in reduced areas due to installation costs, expensive maintenance and impractical use with actual machines used along the growth period (sowing, mechanical weeding, cutting and threshing process). Reduced tillage also has been implemented in small areas of Durango, but then abandoned due to specialized machinery requirements, increment in weed populations and excessive herbicide use; as well as long periods required before registering yield increments.

The use of the biological pests control has been implemented releasing natural predators of plague-insects causing production problems, but no clear results were observed. Cultivar choice is present in common bean including several landraces (Negro San Luis, Bayo Rata, Canario), improved cultivars (Pinto Saltillo, Pinto Centauro, PID 1 and NOD 1) and breeding lines (PT14053, NGO14013), although low genetic variation has been observed for resistance to *Fusarium* and other fungi included in the root-rot complex. However, differences have been observed for plant surviving or escape strategies avoiding severe problems caused by root and aerial plant diseases.

In chili peppers difficulties have been observed for cultivar change, due to traditional use of specific open-pollinated cultivars (landraces), high seed prices for commercial hybrids and specific traits observed in landraces for fresh and dried fruits, as well as for processed fruit (chile pasado) flavor. Similar traits were considered in other chili pepper cultivars (puya, güerito) used for specific preparations included in Durango's cousine (chile con queso and frijoles charros). However outstanding results have been observed by using tomato 'big plant', produced under nursery conditions [50], and technology could be used for producing pathogen-free chili seedlings.

4. Conclusions

The *Fusarium* genus causes significant reductions in yield and seed or fruits quality in common bean, chili pepper and other crops sown in North-Central México. Low income for farmers and total crop losses are also observed in *Fusarium* infested plots affecting food availability and the local economy. Modern agricultural practices should be validated and implemented for sustainable production in com-mon bean, chili and other important crops used in the Mexican highlands. Breeding for plant adaptation, disease resistance, water productivity and product quality are the main concepts in modern and sustainable agriculture.

Author details

Julio César Ríos Saucedo[1], María Gabriela Ramírez-Valadez[2], Saúl Santana Espinoza[3], Maihualy Martínez-Fernández[2] and Rigoberto Rosales-Serna[4*]

1 Sustainable Forest Management and Environmental Services, INIFAP - Campo Experimental Valle del Guadiana, Durango, México

2 Horticulture Program, INIFAP - Campo Experimental Valle del Guadiana, Durango, México

3 Forest and Agriculture Health Program, INIFAP - Campo Experimental Valle del Guadiana, Durango, México

4 Common Bean and Chickpea Program, INIFAP - Campo Experimental Valle del Guadiana, Durango, México

*Address all correspondence to: rosales.rigoberto@inifap.gob.mx

References

[1] Sánchez AJH. Etiología y daños de las pudriciones radicales del frijol (*Phaseolus vulgaris*) en el estado de Durango. México: Tesis de I. A. Universidad Autónoma de Chapingo; 1983. p. 66

[2] Apodaca SMA, Zavaleta ME, Osaka KS, García ER, Valenzuela UJG. Hospedantes asintomáticos de *Fusarium oxysporum* Schlechtend. f. sp. *radicis-lycopersici* W. R. Jarvis y Shoemaker en Sinaloa, México. Revista Mexicana de Fitopatología. 2004;**22**(1):7-13

[3] Valenciano JB, Casquero PA, Boto JA, Marcelo V. Evaluation of occurrence of root rots on bean plants (*Phaseolus vulgaris*) using different sowing methods and with different techniques of pesticide application. New Zealand Journal of Crop and Horticultural Science. 2006;**34**(4):291-298

[4] Bodah ET. Root rot diseases in plants: A review of common causal agents and managemwnt strategies. Agricultural Research & Technology: Open Accesss Journal. 2017;**5**(3):0056-0063

[5] Jiménez GJC, Acosta GJA. Effect of crop density on yield of bean Pinto Saltillo under irrigation in Chihuahua, Mexico. Revista Mexicana de las Ciencias Agrícolas. 2013;**4**(2):243-257

[6] Martínez FM, Velázquez VR. Especies de mosquita blanca presentes en el área de Poanas, Durango. Durango México: Folleto Técnico Núm.118. INIFAP-CIRNOC-Campo Experimental Valle del Guadiana; 2020. 25 p

[7] Pandey P, Irulappan V, Bagavathiannan MV, Senthil-Kumar M. Impact of combined abiotic and biotic stresses on plant growth and avenues for crop improvement by exploiting physio-morphological trais. Frontiers in Plant Science. 2017;**8**:537

[8] Mohanapriya R, Naveenkumar R, Balabaskar P. Survey, virulence and pathogenicity of root rot incidence of cowpea in selected districts of Tamilnadu cused by *Macrophomina phaseolina* (Tassi.) Goid. International Journal of Current Microbiology and Applied Sciences. 2017;**6**(3):694-705

[9] CIAT (Centro Internacional de Agricultura Tropical). Stages of development of the common bean plant; Study guide to be used as supplement to auditorial unit on the same topic. Colombia: Cali; 1986. 32 p

[10] Lira MK, Hernández RMD, Martínez MC, Rosales SR, Mayek PN, González PJM. Root rot fungi associated to common beans in Durango, México. Annual Report of the Bean Improvement Cooperative (BIC). 2008;**51**:224-225

[11] Padilla RJS, Ochoa MR, Rosales SR, Ibarra PFJ, Méndez RA, Hernández DS, et al. Analysis of *Fusarium*-common beans pathosystem in Aguascalientes, México. In: Askun T, editor. *Fusarium*. London, U. K: IntechOpen; 2018. pp. 101-117

[12] Chew MYI, Vega PA, Palomo RM, Jiménez DF. Principales enfermedades del chile (*Capsicum annuum* L.). Matamoros, Coah. Méx: Folleto Técnico Núm. 15. INIFAP-CIRNOC-Campo Experimental La Laguna; 2008. 32 p

[13] Anaya LJL, González CMM, Villordo PE, Rodríguez GR, Rodríguez MR, Guevara GRG, et al. Selección de genotipos de chile resistentes al complejo patógénico de la marchitez. Revista Mexicana de Ciencias Agrícolas. 2011;**2**(3):373-383

[14] Balodi R, Bisht S, Ghatak A, Rao KH. Plant disease diagnosis: technological advancements and challenges. Indian Phytopathology. 2017;**70**(3):275-281

[15] Ntatsi G, Karkanis A, Tran F, Savvas D, Iannetta PPM. Which agronomic practices increase the yield and quality of common bean (*Phaseolus vulgaris* L.)? A systematic review protocol. Agronomy. 2020;**10**(7):1-10

[16] SIAP (Servicio de Información Agroalimentaria y Pesquera). Anuario estadístico de la producción agrícola. México; 2021 Consulted online 22/06/2021. Available from: https://nube.siap.gob.mx/cierreagricola/

[17] Tijerina CAD, Torres HD, Reyes JI, Domínguez MPA, Jiménez OR, Rosales SR. Agenda técnica agrícola Durango y La Laguna Compilación. México: SAGARPA-COFUPRO-INIFAP; 2017. 196 p

[18] Akwa TE, Maingi JM, Birgen JK. Characterisation of fungi of stored common bean cultivars grown in Menoua division, Cameroon. Journal of Plant Physiology and Pathology. 2020;**9**:1

[19] Musoni A, Kimani P, Narla RD, Buruchara R, Kelly JD. Inheritance of *Fusarium* wilt (*Fusarium oxysporum* F. sp. *phaseoli*) resistance in climbing beans. African Journal of Agricultural Research. 2010;**5**(5):399-404

[20] Singh SP, Gepts P, Debouck DG. Races of common bean (*Phaseolus vulgaris*, Fabaceae). Economic Botany. 1991;**45**(3):379-396

[21] Haus MJ, Pierz LD, Jacobs JL, Wiersma AT, Awale HE, Chilvers MI, et al. Preliminary evaluation of wild bean (*Phaseolus* spp.) germplasm to *Fusarium cuneirostrum* and *Fusarium oxysporum*. Crop Science. 2021;**2021**:1-11

[22] Schneider KA, Grafton KF, Kelly JD. QTL analysis of resistance to *Fusarium* root rot in bean. Crop Science. 2001; **41**(2):535-542

[23] Zavaglia-Pereira MJ, Patto-Ramalho MA, de Barbosa-Abreu AF. Iheritance of resistance to *Fusarium oxisporum* f. sp. *phaseoli* Brazilian race in common bean. Scientia Agricola. 2009;**66**(6):788-792

[24] Lutz MP, Feichtinger G, Défago G, Duffy B. Mycotoxigenic *Fusarium* and deoxynivalenol production repress chitinase gene expression in the biocontrol agent *Trichoderma atroviride* P1. Applied and Environmental Microbiology. 2003;**69**:3077-3084

[25] Drakopoulos D, Meca G, Torrijos R, Marty A, Kägi A, Jenny E, et al. Control of *Fusarium graminearum* in wheat with mustard-based botanicals: from *in vitro* to *in planta*. Frontiers in Microbiology. 2020;**11**:1595

[26] Rosales SR. Producción de semilla calificada para incrementar el rendimiento y calidad del frijol producido en Durango. Durango Dgo., México: Desplegable para Productores Núm. 102. INIFAP-CIRNOC-Campo Experimental Valle del Guadiana; 2018. 2 p

[27] SNICS (Servicio Nacional de Inspección y Certificación de Semillas). Regla para la calificación de semillas. Frijol (*Phaseolus vulgaris* L.). México: CDMX; 2021. 23 p

[28] Cruz CE, Acosta GJA, Reyes ML, Cueto WJA. Variedades de frijol (*Phaseolus vulgaris* L.) del INIFAP. Ciudad de México: INIFAP-Oficinas Centrales. Libro Técnico Núm. 2; 2021. 98 p

[29] Ferreira TF, Carvalho MV, de Ferreira VF, Mavaieie PR, Guimarães GC, Oliveira JA. Sanitary quality of soybean seeds treated with fungicides and insecticides before and after storage. Journal of Seed Sciences. 2019;**41**(3):293-300

[30] Sales FJ, Pinto JEBP, de Oliveira JA, Botrel PP, Silva FG, Corrêa RM. The germination of bush mint (*Hyptis marrubioides* EPL.) seeds as a function of harvest stage, light, temperature and

duration of storage. *Acta Scientiarum*, Agronomy. 2011;**33**(4):709-713

[31] Kumar R, Gupta A, Srivastava S, Devi G, Singh VK, Goswami SK, Gurjar MS, Aggarwal R. Diagnosis and detection of seed-borne fungal phytopathogens. In: Kumar R, Gupta A editors. Seed-borne diseases of agricultural crops: Detection, diagnosis & management. Singapore: Springer; 2020. pp. 107-142

[32] Gupta S, Dubey A, Singh T. *Fusarium semitectum* as a dominant seed-borne pathogen in *Dalbergia sissoo* Roxb., its location in seed and its phytopathological effects. Indian Journal of Fundamental and Applied Life Sciences. 2011;**1**(1):5-10

[33] Tsedaley B. Review on seed health tests and detection methods of seedborne diseases. Bilogy Agriculture and Healthcare. 2015;**5**(5):176-184

[34] Vega GTA, López UGA, Allende MR, Amarillas BLA, Romero GSJ, López OCA. 2019. Aggressiveness and molecular characterization of *Fusarium* spp. associated with foot rot and wilt in tomato in Sinaloa, Mexico. 3 Biotech 9,
276.

[35] Aguirre MCL, Iturriaga FG, Ramírez PJG, Covarrubias PJ, Chablé MF, Raya PJC. El chile (*C. annuum* L.), cultivo y producción de semilla. Ciencia y Tecnología Agropecuaria de México.
2017;**5**(1):19-31

[36] Chigoziri E, Ekefan EJ. Seed borne fungi of chilli pepper (*Capsicum frutescens*) from pepper producing areas of Benue State, Nigeria. Agriculture and Biology Journal of North America. 2013;**4**(4):370-374

[37] Chávez SAL, Inzunza IMA, Mendoza MSF, Sánchez CI, Román LA. Producción de chile jalapeño (*Capsicum annum* L.) con diferentes tipos de acolchado plástico y riego por goteo-cintilla. Revista Chapingo Serie Zonas Áridas. 2007;**VI**(1):67-75

[38] Rivera CJM, Brown JK, Melgar MJC, Weller S. Manchas foliares de tomate y chile causadas por bacterias: Su reconocimiento y manejo integrado. La Lima, Cortés, Honduras: FHIA-USAID-IPM/CRSP; 2014. 2 p

[39] Hernández CFD, Lira SRH, Gallegos MG, Hernández SM, Solis GS. Biocontrol de la marchitez del chile con tres especies de *Bacillus* y su efecto en el crecimiento y rendimiento. ΦYTON. 2014;**83**:49-55

[40] Suryanto D, Patonah S, Munir E. Control of *Fusarium* wilt of chili chitinolytic bacteria. HAYATI Journal of Biosciences. 2010;**17**(1):5-8

[41] Nicolaisen M, Supronienė S, Nielsen LK, Lazzaro I, Spliid NH, Justesen AF. Real-time PCR for quantification of eleven individual *Fusarium* species in cereals. Journal of Microbiological Methods. 2009;**76**(3) :234-240

[42] Velázquez VMG, Bautista BS, Hernández LN. Estrategias de control de *Rhizopus* stolonifer Ehrenb. (Ex Fr.) Lind, agente causal de pudriciones postcosecha en productos agrícolas. Revista Mexicana de Fitopatología. 2008;**26**(1):49-55

[43] Mukuma C. Morphological and molecular identification and characterization of dry bean fungal root rot pathogens in Zambia. In: Theses, Dissertations, and Student Research in Agronomy and Horticulture. USA: University of Nebraska-Lincoln; 2016. 112 p

[44] Cifuentes RD. Prácticas de patología vegetal. EDITUM, Ediciones de la Universidad de Murcia. Murcia, España: Campus de Espinardo; 1990. 52 p

[45] Gil OR, Palazón EC, Cuartero ZJ. Genetics of resistance to *Phytophthora capsici* in México. The Netherlands: Synopses of the IV meeting of the *Capsicum* working group of Eucarpia. I. V. T. Wageningen; 1991. pp. 52-56

[46] Méndez AR, Rodríguez GR, Ramírez MM, Álvarez OMG, Vázquez GE, Cavazos GA, et al. Identificación de fuentes de resistencia a pudriciones de la raíz en germoplasma de chile serrano (*Capsicum annuum* L.). Revista Mexicana de Ciencias Agrícolas. 2015;**6**(7):1507-1518

[47] Wezel A, Casagrande M, Celette F, Vian J-F, Ferrer A, Peigné J. Agro-ecological practices for sustainable agriculture A review Agronomy for Sustainable Development. 2014;**34**:1-20

[48] Ríos SJC, Rosales SR, Escobedo RI, Gutiérrez SJV, Nava BCA, Fernández MM, Domínguez MPA, Santana ES. Calidad y carga microbiana de la biomasa producida con especies vegetales cultivadas de forma intensiva. Agrofaz-Journal of environmental and Agroecological Sciences. 2019;**1**(1):43-55

[49] Borja BM, Rosales SR, Sigala RMA, Sarmiento LH, Rosales MS. Eficiencia productiva y económica de sistemas agroforestales pino y cultivos anuales en el estado de Durango. Pabellón de Arteaga, Ags., México: Folleto Técnico Núm 72. INIFAP-CIRNOC-Campo Experimental Pabellón; 2016. 31 p

[50] Huchín AS, Reveles HM, Merlín BE, Trejo CR, Galindo RMA, Cisneros ROB, et al. Uso de diferentes sustratos en la producción de plántulas de edad avanzada (Big Plant) de tomate (*Licopersicum esculentum* Mill.) en invernadero. Memoria de la XXIII Semana Internacional de Agronomía FAZ-UJED; 2011;**23**:892-895

Permissions

All chapters in this book were first published by InTech Open; hereby published with permission under the Creative Commons Attribution License or equivalent. Every chapter published in this book has been scrutinized by our experts. Their significance has been extensively debated. The topics covered herein carry significant findings which will fuel the growth of the discipline. They may even be implemented as practical applications or may be referred to as a beginning point for another development.

The contributors of this book come from diverse backgrounds, making this book a truly international effort. This book will bring forth new frontiers with its revolutionizing research information and detailed analysis of the nascent developments around the world.

We would like to thank all the contributing authors for lending their expertise to make the book truly unique. They have played a crucial role in the development of this book. Without their invaluable contributions this book wouldn't have been possible. They have made vital efforts to compile up to date information on the varied aspects of this subject to make this book a valuable addition to the collection of many professionals and students.

This book was conceptualized with the vision of imparting up-to-date information and advanced data in this field. To ensure the same, a matchless editorial board was set up. Every individual on the board went through rigorous rounds of assessment to prove their worth. After which they invested a large part of their time researching and compiling the most relevant data for our readers.

The editorial board has been involved in producing this book since its inception. They have spent rigorous hours researching and exploring the diverse topics which have resulted in the successful publishing of this book. They have passed on their knowledge of decades through this book. To expedite this challenging task, the publisher supported the team at every step. A small team of assistant editors was also appointed to further simplify the editing procedure and attain best results for the readers.

Apart from the editorial board, the designing team has also invested a significant amount of their time in understanding the subject and creating the most relevant covers. They scrutinized every image to scout for the most suitable representation of the subject and create an appropriate cover for the book.

The publishing team has been an ardent support to the editorial, designing and production team. Their endless efforts to recruit the best for this project, has resulted in the accomplishment of this book. They are a veteran in the field of academics and their pool of knowledge is as vast as their experience in printing. Their expertise and guidance has proved useful at every step. Their uncompromising quality standards have made this book an exceptional effort. Their encouragement from time to time has been an inspiration for everyone.

The publisher and the editorial board hope that this book will prove to be a valuable piece of knowledge for researchers, students, practitioners and scholars across the globe.

List of Contributors

Olawale Arogundade, Titilayo Ajose, Itinu Osijo, Hilary Onyeanusi and Joshua Matthew
National Horticultural Research Institute (NIHORT), Ibadan, Oyo State, Nigeria

Taye H. Aliyu
Department of Crop Protection, University of Ilorin, Ilorin, Kwara State, Nigeria

Mithun Rudrapal
Sandip Institute of Pharmaceutical Sciences, Nashik, India

Khomendra Kumar Sarwa
Department of Pharmacy, Govt. Girl's Polytechnic, Raipur, India

S.R. Krishna Motukuri and Nallamothu Jaswanthi
Department of Biotechnology, Koneru Lakshmaiah Education Foundation, Guntur, Andhra Pradesh, India

Balasubramanian Arul and Ramalingam Kothai
Vinayaka Mission's College of Pharmacy, Vinayaka Mission's Research Foundation (Deemed to be University), Salem, Tamil Nadu, India

Eme Efiom Osim
Department of Physiology, University of Calabar, Calabar, Nigeria
Department of Physiology, Abia State University, Uturu, Nigeria

Harshita Singh
Department of Vegetable Science, Chaudhary Charan Singh Haryana Agricultural University, Hisar, Haryana, India

Narender K. Bharat
Department of Seed Science and Technology, Dr. Y.S. Parmar University of Horticulture and Forestry, Nauni Solan, HP, India

Cristofora Hanny Wijaya and Bunga Rana
IPB University, Bogor, Indonesia

Manik Harda
Gadjah Mada University, Yogyakarta, Indonesia

Nilton Oliveira Junior and Valter Padulla
Condor Non-Lethal Technology S/A, Rio de Janeiro, Brazil

Vinnicius Ferraço and Giovana Ferronato
Instituto Senai de Inovação em Química Verde, Rio de Janeiro, Braziltt

Raj Kiran, Jameel Akhtar, Pardeep Kumar and Meena Shekhar
Division of Plant Quarantine, ICAR-National Bureau of Plant Genetic Resources, New Delhi, India

Udemeobong Okon and Joshua Iwasam
Department of Physiology, University of Calabar, Calabar, Nigeria

Suryapal Singh
Department of Seed Science and Technology, Chaudhary Charan Singh Haryana Agricultural University, Hisar, Haryana, India

Anyauba Nmaju and Azubuike Nwankwo
Department of Physiology, Abia State University, Uturu, Nigeria

Lorena Barra-Bucarei
Instituto de Investigaciones Agropecuarias, Chillán, Chile

Javiera Ortiz
Universidad de Concepción, Chillán, Chile

Javier Echave, María Carpena, Miguel Ángel Prieto and Jesús Simal-Gandara
Nutrition and Bromatology Group, Analytical and Food Chemistry Department, Faculty of Food Science and Technology, University of Vigo, Ourense, Spain

Antía G. Pereira
Nutrition and Bromatology Group, Analytical and Food Chemistry Department, Faculty of Food Science and Technology, University of Vigo, Ourense, Spain
Centro de Investigação de Montanha (CIMO), Instituto Politécnico de Bragança, Bragança, Portugal

Samia Ben Mansour-Gueddes and Dhouha Saidana-Naija
Olive Tree Institute, Sousse, Tunisia

Saúl Santana Espinoza
Forest and Agriculture Health Program, INIFAP - Campo Experimental Valle del Guadiana, Durango, México

Julio César Ríos Saucedo
Sustainable Forest Management and Environmental Services, INIFAP - Campo Experimental Valle del Guadiana, Durango, México

María Gabriela Ramírez-Valadez and Maihualy Martínez-Fernández
Horticulture Program, INIFAP - Campo Experimental Valle del Guadiana, Durango, México

Rigoberto Rosales-Serna
Common Bean and Chickpea Program, INIFAP - Campo Experimental Valle del Guadiana, Durango, México

Index